T0269174

# MARS: AN INTRODUCTION TO ITS INTERIOR, SURFACE AND ATMOSPHERE

Our knowledge of Mars has changed dramatically in the past 40 years due to the wealth of information provided by Earth-based and orbiting telescopes, and spacecraft investigations. Recent observations suggest that water has played a major role in the climatic and geologic history of the planet. This textbook covers our current understanding of the planet's formation, geology, atmosphere, interior, surface properties, and potential for life.

This interdisciplinary textbook encompasses the fields of geology, chemistry, atmospheric sciences, geophysics, and astronomy. Each chapter introduces the necessary background information to help the non-specialist understand the topics explored. It includes results from missions through 2006, including the latest insights from Mars Express and the Mars Exploration Rovers.

Containing the most up-to-date information on Mars, this textbook is essential reading for graduate courses and an important reference for researchers.

NADINE BARLOW is Associate Professor in the Department of Physics and Astronomy at Northern Arizona University. Her research focuses on Martian impact craters and what they can tell us about the distribution of subsurface water and ice reservoirs.

# Cambridge Planetary Science

Series Editors: Fran Bagenal, David Jewitt, Carl Murray, Jim Bell, Ralph Lorenz, Francis Nimmo, Sara Russell

*Books in the series*

[†] Issued as a paperback

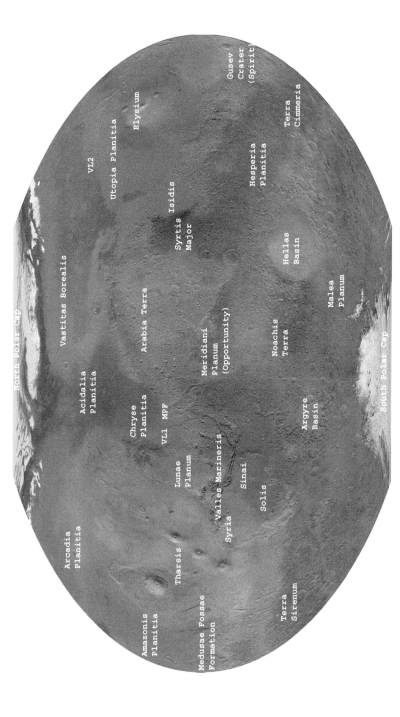

Shaded relief map of Mars showing the locations of major features discussed in the text. Landing sites of the two Viking Landers (VL1 and VL2), Mars Pathfinder (MPF), and the two Mars Exploration Rovers (Spirit and Opportunity) are also shown. (Shaded relief map courtesy of National Geographic Society/MOLA Science Team/Malin Space Science Systems (MSSS)/NASA/JPL)

# MARS: AN INTRODUCTION TO ITS INTERIOR, SURFACE AND ATMOSPHERE

## NADINE BARLOW
*Northern Arizona University*

CAMBRIDGE
UNIVERSITY PRESS

CAMBRIDGE
UNIVERSITY PRESS

University Printing House, Cambridge CB2 8BS, United Kingdom

Published in the United States of America by Cambridge University Press, New York

Cambridge University Press is part of the University of Cambridge.

It furthers the University's mission by disseminating knowledge in the pursuit of
education, learning and research at the highest international levels of excellence.

www.cambridge.org
Information on this title: www.cambridge.org/9781107644878

© N. Barlow 2008

This publication is in copyright. Subject to statutory exception
and to the provisions of relevant collective licensing agreements,
no reproduction of any part may take place without the written
permission of Cambridge University Press.

First published 2008
First paperback edition 2014

*A catalogue record for this publication is available from the British Library*

ISBN 978-0-521-85226-5 Hardback
ISBN 978-1-107-64487-8 Paperback

Cambridge University Press has no responsibility for the persistence or accuracy of
URLs for external or third-party internet websites referred to in this publication,
and does not guarantee that any content on such websites is, or will remain, accurate
or appropriate.

# Contents

*The color plates are situated between pages 84 and 85*

# Preface

It is an exciting time to be a planetary scientist specializing in Mars research. I have had the privilege of experiencing our changing views of Mars since the beginning of space missions to our neighbor. Mariners 6 and 7 flew by the planet shortly after I became interested in astronomy at age 10. I checked the news every day when Mariner 9 began to reveal the geologic diversity of Mars. The Viking missions started their explorations as I was entering college and the Viking 1 lander ceased operations just as I was starting to utilize the orbiter data in my Ph.D. thesis. Over the subsequent years I grieved the lost missions and cheered the successful ones. I feel extremely fortunate to be able to work in such an exciting field and contribute to our expanding knowledge of the planet and its history.

I have taught graduate courses about Mars at the University of Houston Clear Lake, University of Central Florida, and Northern Arizona University. The 1992 University of Arizona Press book *Mars* is the best compilation of our knowledge through the Viking missions, but has become increasingly deficient as Mars Pathfinder, Mars Global Surveyor, Mars Odyssey, Mars Express, and the Mars Exploration Rovers have revealed new facets of Mars' evolution. A few years ago I developed a course pack for my students which I updated prior to each term when I taught the course. This book is an expanded version of that course pack and is appropriate for graduate students in planetary science, professional scientists, and advanced undergraduate science majors.

*Mars: An Introduction to its Interior, Surface and Atmosphere* focuses on what we have learned about Mars since 1992. I have had no delusions that I could replicate the excellent detailed summaries provided by the experts in the 1992 *Mars* book. What I have endeavored to do is to expand upon that treatise by summarizing the latest discoveries and how they are once again changing our paradigm of Mars.

The study of Mars is a very interdisciplinary science, covering sciences as diverse as geology, geophysics, geochemistry, atmospheric dynamics, and biology.

Mars researchers specialize in a specific discipline and can have difficulty understanding the other aspects of Mars studies. I have taught my graduate level Mars class to geology, physics, astronomy, and engineering students and have seen how they struggle with material outside of their discipline. I have therefore structured this book to include enough background material so someone from another discipline can understand and appreciate the advances made in other fields. I assume the reader has sufficient background to understand basic geology terms, calculus-based introductory physics, and mathematics through differential equations. However, it is impossible to provide detailed coverage of every aspect of Mars research in such a book. I have therefore provided an extensive reference list and encourage the reader whose interest is piqued by a particular topic to explore the original literature. I apologize in advance to my colleagues who find some of their works not included in the references. The Mars literature is voluminous and I have attempted to provide a reasonable sampling of the articles and books which cover the currently accepted views of the planet and some of the ongoing debates.

It is difficult to write a book like this while so many Mars missions are still operating. Several times I thought I had completed a specific chapter only to go back later and revise it as new discoveries were announced. Therefore the reader needs to recognize that the book includes information only through the end of 2006. I fully expect that results from the recently arrived Mars Reconnaissance Orbiter together with new discoveries by Odyssey, Mars Express, Spirit, and Opportunity will make some of the discussion in this book obsolete very soon. We are obviously in the golden age of Mars exploration, but now is a good time to step back and summarize the dramatic shift in our view of Mars produced during the past ten years of spacecraft and telescopic observations.

This book would not have been possible without the input and support of many people. My mother Marcella, my late father Nathan, and my sister Lynn have provided love and support throughout the years, even when they probably thought I was crazy for being such a Mars fanatic. I also appreciate all my friends and family who are not scientists but who have always shown great interest in the work that I do. I have had many wonderful professors who have inspired me and would like to specifically acknowledge the mentorship of James Pesavento and Robert Strom. My colleagues at Northern Arizona University have been a great support system and I especially acknowledge Dean Laura Huenneke and chairs Tim Porter and David Cornelison who encouraged me to undertake this project even when I was not yet tenured. The staff at Cambridge University Press has been a joy to work with and I want to thank Helen Goldrein for her patience and encouragement. Finally, this book would not have been possible without the work of my friends and colleagues in the Mars community. Here's to many more years of exciting discoveries!

# 1

## Introduction to Mars

### 1.1 Historical observations

#### 1.1.1 Pre-telescopic observations

Mars has been the focus of intense scientific interest and study throughout recorded history. Even before the advent of the telescope in 1609, astronomers carefully charted the motion of Mars across the sky. The planet's obviously reddish-orange color led many ancient civilizations to name the planet after war or warrior gods. Our current use of the name Mars comes from the Roman God of War. Large martian sinuous valleys (*vallis*) are named after the term for Mars in different languages: hence Ares Vallis (Greek name for Mars), Augakuh Vallis (Incan), and Nirgal Vallis (Babylonian).

Careful observations of Mars' motion across the celestial sphere led early astronomers to deduce two things about the planet. First they determined that Mars' sidereal period (time to return to same position relative to the stars) is about 687 Earth days (1.88 Earth years). The Polish astronomer Nicolaus Copernicus found that the sidereal period ($P$) of a planet located beyond the Earth's orbit is related to its synodic period ($S$; time for planet to return to same Earth–Sun–planet configuration) by

$$\frac{1}{P} = 1 - \frac{1}{S}.$$ 
(1.1)

Using this relationship, we can determine that the synodic period of Mars is 2.14 Earth years.

The second thing that pre-telescopic observers noticed about Mars was its strange looping path across the sky. While planets tend to slowly travel from west to east across the background of stars over the period of several nights, occasionally they reverse course and travel east to west for a period of time before resuming their normal west-to-east motion. This retrograde (east-to-west) motion is most notice-able for planets closest to the Earth, and thus Mars' retrograde motion is apparent even to naked-eye observers. The geocentric model of the universe had extreme difficulty explaining retrograde motion, requiring the use of hundreds of small circles upon orbital circles (epicycles and deferents). However, retrograde motion was easily explained when Copernicus rearranged the view of the Solar System in 1543 by placing the Sun at the center and having Earth orbit the Sun along with the other planets. In the heliocentric model, retrograde motion results when one planet catches up to and overtakes another during their orbital motions.

Mars also played a major role in determining the shapes of planetary orbits. It was Tycho Brahe's very accurate and voluminous observations of Mars' celestial positions that led Johannes Kepler in 1609 to deduce that planetary orbits were elliptical with the Sun at one focus of the orbit. Mars has the second most elliptical orbit of the eight major planets in the Solar System – Mercury's orbital eccentricity is higher but the planet is difficult to observe due to its proximity to the Sun.

### 1.1.2 Telescopic observations from Earth and space

Although Galileo's small telescope was unable to reveal anything other than the reddish-orange disk of the planet in 1609, larger telescopes slowly coaxed more information from the planet. By 1610, Galileo reported that Mars can show a gibbous phase, which subsequent observers verified. The first report of albedo markings on the surface was published in 1659 by Christiaan Huygens, whose map showed a dark spot which was likely Syrtis Major. The identification of surface albedo markings allowed astronomers to determine that the rotation period of Mars was approximately 24 hours. The bright polar caps apparently were not noticed until Giovanni Cassini reported them in 1666. Cassini's nephew, Giacomo Maraldi, made detailed observations of the polar caps during several oppositions, including the favorable opposition of 1719. Among his discoveries were that the south polar cap was not centered on the rotation pole, that the polar caps and equatorial dark areas displayed temporal variations, and that a dark band occurs around the edge of the receding polar cap (which he interpreted as meltwater).

Sir William Herschel observed Mars from 1777 to 1783 and was the first to determine that Mars' rotation axis was tipped approximately 30° from the perpendicular to its orbit. This result showed that Mars experiences four seasons, similar to the Earth. Herschel also determined the planet's rotation period to be 24 hours 39 minutes 21.67 seconds. Herschel deduced the presence of a thin atmosphere around Mars based on the changes he saw in the appearance of the planet, which he attributed to clouds. These were primarily the white clouds which are now known to be composed of ice particles. The yellow dust clouds were first reported by Honoré Flaugergues in 1809.

Major advances in our understanding of Mars began in 1830 during a close approach between Mars and Earth. The first complete map of Mars was derived from observations during this time and published in 1840 by Johan von Mädler and Wilhelm Beer – this was the first map to establish a latitude–longitude system for the planet, with the zero longitude line defined through a small, very dark spot. They also refined the rotation period of Mars to 24 hours 37 minutes 22.6 seconds (within 0.1 second of the currently accepted value). Numerous drawings of Mars were made between 1830 and the early twentieth century and these drawings were gradually incorporated into maps by William Dawes in 1864, Richard Proctor in 1867, Nicolas Flammarion (1876), and E.M. Antoniadi (between 1901 and 1930). Although Proctor and Flammarion both named features on their maps, the nomenclature system currently used for martian features is based on one proposed by Giovanni Schiaparelli on his 1877 map.

Mars and Earth were very close in 1877, resulting in a surge of new discoveries. Principal among these was the discovery of Mars' two small moons, Phobos and Deimos, by Asaph Hall. High-altitude clouds were detected as white spots along the morning and evening limbs of the planet by Nathaniel Green and the first attempts to photograph the planet were also made this year by M. Gould. But it was another observation made in 1877 that would focus considerable attention on Mars for many years to come: Schiaparelli's observations of thin dark lines crossing the martian surface, which he called *canali*.

Schiaparelli reported thin dark lines crossing the martian surface, but he was unsure of their origin. As a result, he used the generic term "channel" to describe these features. A channel is a natural feature which can be formed by flowing liquid/ ice, tectonics, or wind. The Italian word for channel is "canali," which unfortunately was mistranslated into English as "canal," a word that implies a waterway con- structed by intelligent beings.

The discovery of these "canals" simply augmented several other observations which people felt supported the idea of life on Mars. Mars displays a number of Earth-like characteristics which were already known in the nineteenth century. Mars' rotation period is about 37 minutes longer than an Earth day, and, due to the tilt of its rotation axis, it undergoes four seasons just like the Earth. Telescopic observations had revealed the presence of polar caps and an atmosphere, although their compositions were unknown. But one of the most intriguing observations for possible life on Mars was the "wave of darkening." Telescopic observations revealed that as one hemisphere's polar cap began to recede in the spring, the region immediately surrounding the polar cap became noticeably darker. As the polar cap continued to recede into summer, the area of darkening extended towards the equator. As fall arrived and the polar cap began to increase in size, the "wave of darkening" reversed itself and the hemisphere

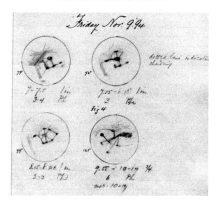

Figure 1.1 Image of the martian canals (dark lines) and putative lakes (round dots), drawn by Percival Lowell on the night of 9 November 1894. (Image courtesy of Lowell Observatory Archives.)

underwent a "wave of brightening" from the equator toward the poles. Most people attributed the wave of darkening to the melting of water ice at the polar caps in the spring and summer and the greening of surface vegetation as it absorbed this water.

Schiaparelli's *canali* were quickly accepted as evidence that not only vegetation but also intelligent life existed on Mars. This idea was promulgated by Percival Lowell, a wealthy Bostonian who founded the Lowell Observatory in Flagstaff, AZ, in 1894 specifically to study the martian canals. Lowell observed hundreds of single and double canals using the 0.6-m Clark refracting telescope at the Observatory (Figure 1.1) and wrote several books describing his thoughts on the origin of these canals. According to Lowell, ancient Mars retained a thicker atmosphere, which led to temperate conditions on the surface, including abundant liquid water. A race of Martians arose under these conditions and settled the entire planet. But since Mars is only 52% the size of the Earth, the atmosphere gradually began to escape to space, cooling the surface and making liquid water less abundant. The Martians moved to the warm equatorial region of the planet and constructed the elaborate network of canals to bring water from the polar regions to the thirsty masses at the equator. Lowell realized that the canals themselves would likely be too small to be resolved with any Earth-based telescope, so he argued that the dark lines he observed were regions of vegetation bordering the canals. Lowell's books and public lectures always drew large, enthusiastic crowds and many science fiction books about Martians resulted from this discussion (e.g., *The Martian Chronicles* and *The War of the Worlds*).

Most astronomers, however, were not convinced that martian canals existed. More powerful telescopes did not show dark lines but dark blotches across the surface. Scientists argued that the linear canals were optical illusions caused by the human mind "connecting the dots" when observing at the limits of a telescope's ability. Such conditions are exacerbated by the turbulent atmospheres of both the Earth and Mars. Testing of human subjects confirmed these arguments. Lowell countered that the excellent seeing at his Flagstaff site allowed him to observe features that other telescopes did not reveal. The controversy continued after Lowell's death in 1916 and the development of large telescopes such as the 5-m telescope at Palomar Observatory in California in 1948. It was only after the advent of spacecraft exploration that astronomers were able to say definitively that canals do not exist on Mars and that the "wave of darkening" simply results from the movement of dust and sand across the planet by seasonal winds.

Advances in telescope size and technology have greatly affected the quality and type of astronomical observations in recent years, and Mars studies have been one of the areas to reap these benefits. Infrared observations of Mars from Earth-based telescopes and the Hubble Space Telescope provided evidence of mineralogical variations across the surface, including the presence of hydrated minerals. The advent of adaptive optics, together with observations from the Hubble Space Telescope, have provided dramatic improvements in resolution and allow study of martian features which previously were only seen by orbiting spacecraft at Mars. Radar observations from ground-based radio telescopes have provided important constraints on surface roughness which have been important in the selection of landing sites for landers and rovers. These ground-based roughness measurements have only recently been surpassed by the acquisition of Mars Orbiter Laser Altimeter (MOLA) data from Mars orbit.

Some people argue that ground-based observations of Mars are no longer needed because of the large number of orbiters and landers currently invading the planet (Section 1.2). Nothing could be further from the truth. Ground-based observations can provide the continuous or near-continuous monitoring of rapidly changing events, such as atmospheric phenomena (including dust storm formation and propagation) and polar cap changes. Due to orbital constraints, orbiting spacecraft cannot continuously monitor one location or event each day and landers/rovers are even more restricted in their observations. The wavelengths of observation also are restricted on spacecraft instrumentation. Hubble Space Telescope observations of Mars are few due to the demand for observing time. Thus, ground-based observations still fill an important niche in our ongoing observations of Mars.

## 1.2 Spacecraft missions

Mars has been a major spacecraft destination ever since the early days of space exploration. This was partly driven by its proximity to Earth, but primarily this

Table 1.1 *Missions to Mars*

| Mission | Country | Launch date | Type of mission | Results |
|---|---|---|---|---|
| [Unnamed] | USSR | 10 Oct 1960 | Flyby | Did not reach Earth orbit |
| [Unnamed] | USSR | 14 Oct 1960 | Flyby | Did not reach Earth orbit |
| [Unnamed] | USSR | 24 Oct 1962 | Flyby | Achieved Earth orbit only |
| Mars 1 | USSR | 1 Nov 1962 | Flyby | Radio failed at $106 \times 10^6$ km |
| [Unnamed] | USSR | 4 Nov 1962 | Flyby | Achieved Earth orbit only |
| Zond 2 | USSR | 30 Oct 1964 | Flyby | Passed Mars but radio failed |
| Mariner 3 | US | 5 Nov 1964 | Flyby | Shroud failed to jettison |
| Mariner 4 | US | 28 Nov 1964 | Flyby | Successfully flew by 14 July 1965 |
| Mariner 6 | US | 24 Feb 1969 | Flyby | Successfully flew by 31 July 1969 |
| Mariner 7 | US | 27 Mar 1969 | Flyby | Successfully flew by 5 Aug 1969 |
| Mariner 8 | US | 8 May 1971 | Orbiter | Failed during launch |
| Kosmos 419 | USSR | 10 May 1971 | Lander | Achieved Earth orbit only |
| Mars 2 | USSR | 10 May 1971 | Orbiter/lander | No useful data; lander failed |
| Mars 3 | USSR | 28 May 1971 | Orbiter/lander | Arrived 3 Dec 1971; some data |
| Mariner 9 | US | 30 May 1971 | Orbiter | In orbit 13 Nov 1971 to 27 Oct 1972 |
| Mars 4 | USSR | 21 July 1973 | Orbiter | Failed; flew past Mars 10 Feb 1974 |
| Mars 5 | USSR | 25 July 1973 | Orbiter | Arrived 12 Feb 1974; lasted a few days |
| Mars 6 | USSR | 5 Aug 1973 | Orbiter/lander | Arrived 12 Mar 1974; little data return |
| Mars 7 | USSR | 9 Aug 1973 | Orbiter/lander | Arrived 9 Mar 1974; little data return |
| Viking 1 | US | 20 Aug 1975 | Orbiter/lander | Orbiter lasted 19 June 1976–7 Aug 1980; lander operated 20 July 1976–13 Nov 1982 |
| Viking 2 | US | 9 Sept 1975 | Orbiter/lander | Orbiter lasted 7 Aug 1976–25 July 1978; lander operated 3 Sept 1976–7 Aug 1980 |

| Name | Country | Date | Type | Status |
|------|---------|------|------|--------|
| Phobos 1 | USSR | 7 July 1988 | Orbiter/lander | Lost en route to Mars |
| Phobos 2 | USSR | 12 July 1988 | Orbiter/lander | Lost March 1989 near Phobos |
| Mars Observer | US | 25 Sept 1992 | Orbiter | Lost just before Mars arrival |
| Mars Global Surveyor | US | 7 Nov 1996 | Orbiter | Operated 12 Sept 1997–2 Nov 2006 |
| Mars 96 | Russia | 16 Nov 1996 | Orbiter/lander | Launch vehicle failed |
| Mars Pathfinder | US | 4 Dec 1996 | Lander/rover | Operated 4 July 1997–27 Sept 1997 |
| Nozomi | Japan | 4 July 1998 | Orbiter | In heliocentric orbit |
| Mars Climate Orbiter | US | 11 Dec 1998 | Orbiter | Lost on arrival 23 Sept 1999 |
| Mars Polar Lander/Deep Space 2 | US | 3 Jan 1999 | Lander/penetrators | Lost on arrival 3 Dec 1999 |
| Mars Odyssey | US | 7 Apr 2001 | Orbiter | Arrived 24 Oct 2004; still operating |
| Mars Express | ESA | 2 June 2003 | Orbiter/lander | Arrived 25 Dec 2003; orbiter operating, lander lost upon landing |
| Mars Exploration Rovers: | | | | |
| Spirit | US | 10 June 2003 | Rover | Arrived 3 Jan 2004; still operating |
| Opportunity | US | 7 July 2003 | Rover | Arrived 25 Jan 2004; still operating |
| Mars Reconnaissance Orbiter | US | 12 Aug 2005 | Orbiter | Arrived 10 Mar 2006; still operating |

interest resulted from the question of whether life could ever have existed on our neighbor. Even today much of the incentive for Mars exploration is driven by questions related to whether the planet could have supported life in the past or even today. NASA's mantra of "Follow the water" focuses on how water has affected the geologic and climatic evolution of the planet and its implications for biologic activity. Although there is considerable interest in exploring Mars, it has not been the easiest place to explore. Approximately two-thirds of all spacecraft missions to date have been partial or complete failures. A list of the missions which have been launched through 2006 is given in Table 1.1. These missions are described in more detail in the following sections.

### *1.2.1 US missions to Mars*

The United States began its spacecraft exploration of Mars in 1964 when it launched the Mariner 3 and 4 spacecraft. Although both missions launched successfully, the solar panels powering Mariner 3 did not deploy and that mission ended up in solar orbit. Mariner 4 became the first successful flyby mission of Mars when it passed within 9920 km of the planet's surface on 14 July 1965. It returned 22 close-up photos, revealing a heavily cratered surface (Figure 1.2). Spacecraft instruments confirmed that Mars is surrounded by an atmosphere primarily composed of carbon dioxide ($CO_2$) and that this atmosphere exerted a surface pressure in the range of 500 to 1000 Pascal (Pa). Mariner 4 also reported the existence of a small intrinsic magnetic field. After its successful flyby of Mars, Mariner 4 went into solar orbit where it remains to this day.

Mariners 6 and 7 expanded upon the discoveries made by Mariner 4. Mariner 6 passed within 3437 km of the planet's equatorial region on 3 July 1969 and Mariner 7 followed on 5 August 1969, passing 3551 km over the south polar region. Mariners 6 and 7 returned over 200 pictures of the martian surface and provided measurements of the surface and atmospheric temperatures, surface molecular composition, and atmospheric pressure.

These three flyby missions returned important new information about Mars, but the amount of surface area covered by their cameras was very small and suggested that Mars was a heavily cratered, geologically dead world. That view changed dramatically when Mariner 9 entered orbit on 24 November 1971. Mariner 9 was one of two orbiters launched in May 1971, but its companion, Mariner 8, failed to reach Earth orbit. Mariner 9, along with its Soviet counterparts Mars 2 and 3 (Section 1.2.2) arrived when Mars was shrouded in a global-wide dust storm. Most of the scientific experiments were delayed because of the dust storm and the spacecraft was reprogrammed to image the two martian moons, Phobos and Deimos. By January 1972, the dust storm began to subside and Mariner 9 resumed

Figure 1.2 View of the martian surface taken by the Mariner 4 spacecraft. Craters can be seen, but little else is discernible in this image. (Image PIA02979, NASA/JPL.)

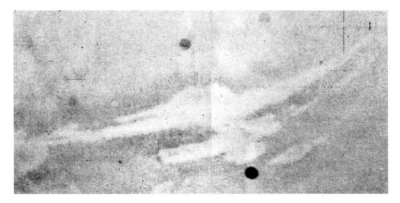

Figure 1.3 One of the first views of the Valles Marineris canyon system, taken by the Mariner 9 spacecraft as the global dust storm was subsiding. (Image PIA02998, NASA/JPL.)

its planned activities. The mission was an outstanding success and operated until 27 October 1972. Among the discoveries made by Mariner 9 were the existence of young volcanoes, canyons (Figure 1.3), and channels in addition to the heavily cratered ancient terrain; meteorological phenomena such as local dust storms, weather fronts, ice clouds, and morning fog; detailed information about the size and shape of Mars, Phobos, and Deimos; temperature gradients within the atmosphere; thermal properties of the surface and atmosphere; and better constraints on the atmospheric composition and pressure.

The discovery of channels formed by a flowing liquid, most likely water, was one of the most exciting results from the Mariner 9 mission. Although Mariners 4, 6, and 7 had unequivocally shown that liquid water cannot exist under the low pressures and temperatures present at the surface today, the channels showed that conditions

Figure 1.4 Viking 1 lander view of the surface of Mars. Viking 1 lander landed in
the Chryse Planitia region within material deposited by flooding through the
outflow channels. Large rock on the left ("Big Joe") is ~2 m high. Rocks and
dunes/drifts are visible in this image. (Image PIA00393, NASA/JPL.)

had likely changed through martian history. With the possibility that liquid water
had existed on the planet's surface, the question of life on Mars again arose. In
response to this resurgence of interest in biology, NASA's next two missions,
Vikings 1 and 2 (Soffen, 1977; Kieffer *et al.*, 1992), included landers which tested
the soil for evidence of microorganisms. In addition to the landers, each spacecraft
also consisted of an orbiter to provide detailed views of the entire planet.

The Viking 1 orbiter/lander spacecraft was launched 20 August 1975 and arrived
at Mars on 19 June 1976. The spacecraft spent approximately one month imaging
the martian surface to find a safe site for the lander. On 20 July 1976 (the seventh
anniversary of Apollo 11 landing on the Moon), the Viking 1 lander set down at
22.48°N 312.03°E in Chryse Planitia within the outwash deposits of several large
channels. The lander operated until 13 November 1982, taking pictures of its
surroundings (Figure 1.4) and testing the soil for evidence of microorganisms. The
orbiter was deactivated on 7 August 1980 when it ran out of attitude-control
propellant.

Viking 2 was launched 9 September 1975 and went into Mars orbit on 7 August
1976. The lander set down on 3 September 1976 in the Utopia Planitia region of
Mars at 47.97°N 134.26°E. The orbiter ran out of attitude-control gas and was
deactivated on 25 July 1978. The lander then used the Viking 1 orbiter as a communication relay and had to be shut down when that orbiter was deactivated on 7
August 1980.

The Viking orbiters mapped the entire surface of Mars and acquired over 52000
images. They provided detailed views of the surface geology and atmospheric
phenomena for about two Mars years, including the first close-up long-term view of
seasonal variations. The Mars Atmospheric Water Detector (MAWD) provided

information on the concentrations and transport of water vapor in the atmosphere and revealed the importance of the seasonal water cycle for both the atmosphere and polar regions (Section 6.3.3). The Infrared Thermal Mapper (IRTM) recorded temperature, albedo, and thermal inertia data for the entire planet, the latter providing important constraints on regional variations in the particle size of surface materials.

The Viking landers measured temperature, density, and composition of the atmosphere as a function of height during their descent. Once on the ground, both landers deployed all of their instruments successfully except for one of the seismometers. Both returned the first color views of the surface of Mars, revealing a rocky and dusty surface. The biology experiments (Section 8.2) were designed to test the soil for organics and detect any possible biogenic waste products resulting from the various tests. The experiments found no unequivocal evidence for life but did discover that the soil readily releases a variety of gases when moistened. Other experiments on the landers included a meteorology station which measured daily pressure, temperature, and wind variations, a magnetic properties investigation that tested the magnetic properties of martian soil, and an investigation which used the robotic sampling arm to determine the physical properties of the soil. Radio signals from the two landers not only helped to identify their exact locations but also have been used to determine accurately the precession and spin deceleration rates for Mars.

Following the end of the Viking missions, US exploration of Mars entered a substantial lull, not broken until the 1990s. The Mars Observer mission, launched on 25 September 1992, was to be a triumphant "return to Mars," with instruments designed to study the surface composition and properties, topography, atmospheric composition and dynamics, and magnetic field environment for at least one entire martian year (Albee *et al.*, 1992). Unfortunately, communication was lost while the spacecraft was entering Mars orbit on 22 August 1993. It is believed that a fuel line cracked during the transfer of fuel in preparation for orbit insertion, causing the spacecraft to tumble out of control.

New instruments have been flown on recent missions in an attempt to recover the science lost on Mars Observer. The first mission to recover some of the initial investigations was Mars Global Surveyor (MGS) (Albee *et al.*, 2001). This was launched on 7 November 1996, successfully achieved Mars orbit on 11 September 1997, and operated until 2 November 2006 when radio communication was lost. The Mars Orbiter Camera (MOC) consisted of three components: a black-and-white narrow-angle camera providing high-resolution imagery (1.5 to 12 m/pixel resolution), and red-and-blue wide-angle cameras which provide context (240 m/pixel resolution) and daily global imaging (7.5 m/pixel resolution). The Mars Orbiter Laser Altimeter (MOLA) used laser pulses to determine topography and roughness

Figure 1.5 Mars Pathfinder view of Chryse Planitia. The rover Sojourner is seen
investigating the large rock "Yogi." "Twin Peaks" are seen in the distance. (Image
PIA01005, NASA/JPL.)

values for the surface. The Thermal Emission Spectrometer (TES) provided
information about the thermal properties of the surface and atmosphere as well as
surface composition. Additional information about the atmosphere and its structure
was obtained through MGS's Radio Science Experiment, which provided daily
weather reports for the planet. The Magnetometer investigation (MAG) provided
the first detailed constraints on an active martian magnetic field and detected
remnant magnetization in surface rocks.

MGS arrived at Mars shortly after the first successful rover was deployed on the
planet. The Mars Pathfinder (MPF) mission was launched 4 December 1996,
the second of NASA's low-cost Discovery missions. The lander set down in the
outwash region of Ares Vallis (19.13°N 326.78°E), a large channel on Mars, on
4 July 1997 (Golombek *et al.*, 1999a) (Figure 1.5). The mission utilized a new type
of landing mechanism, using airbags to cushion the landing of the spacecraft. After
landing, the airbags deflated and retracted, the triangular-shaped lander opened,
and the small rover, called Sojourner (after American civil-rights leader Sojourner
Truth), rolled onto the martian surface. Although the rover was expected to last for
only one week and the lander for one month, they both operated until 27 September
1997. The rover traveled a distance of ~100 m near the lander in its three months
of operation. The lander contained a stereo camera (Imager for Mars Pathfinder:
IMP) and a meteorology package (Atmospheric Structure Instrument/Meteorology
Package: ASI/MET). The rover carried a navigation camera and an Alpha Proton
X-Ray Spectrometer (APXS) to characterize soil and rock compositions. Magnets
attached to the lander also provided information about the magnetic properties
of dust.

The year 1999 was to be an exciting one for Mars exploration with the addition
of two more American spacecraft. The Mars Climate Orbiter (MCO) was launched

on 11 December 1998, and was designed to serve as a martian weather satellite. The mission carried two instruments: an atmospheric sounder (a similar instrument was lost on the Mars Observer mission) and a color imager. Unfortunately an error in converting between English and metric measurement units caused the spacecraft to enter the atmosphere too low during aerobraking operations and the spacecraft probably burned up in the atmosphere. MCO's 1999 companion spacecraft, the Mars Polar Lander (MPL), was to land on the edge of the south polar cap. It included a sample arm which was to dig into the soil and measure water content and thermal properties. Attached to MPL were two small penetrators, the Deep Space 2 (DS2) mission, which were to separate from the spacecraft during descent and impact into the ice-rich materials near the south pole. No communications from either MPL or DS2 were ever received. The MPL landing may have failed due to sensors on the lander's legs shutting off the rockets too soon, causing the spacecraft to crash.

The loss of both 1999 missions was a major setback to America's exploration of Mars. A planned lander for 2001 was canceled because its landing gear was identical to that of MPL – until the loss of MPL was understood and corrected, nobody was willing to spend the money on another mission which could suffer the same fate. The orbiter planned for 2001 proceeded and was renamed the Mars Odyssey mission (in reference to Arthur C. Clarke's book *2001: A Space Odyssey*). Odyssey was to carry the last of the Mars Observer replacement instruments: a Gamma Ray Spectrometer (GRS) instrument package to measure gamma rays and neutrons produced by surface materials either through natural radioactive decay or by interaction with cosmic rays. Odyssey was launched 7 April 2001 and entered Mars orbit on 24 October 2001. In addition to GRS, Odyssey also carries the Thermal Emission Imaging System (THEMIS) instrument which observes the planet in both visible (VIS: 18 m/pixel resolution) and infrared (IR: 100 m/pixel resolution) wavelengths. The IR camera operates during both day and night, providing extremely valuable information about the thermophysical properties of surface materials. Daytime IR also is used to determine the mineralogical variations which occur across the martian surface. GRS constrains abundances of a variety of elements, including water ($H_2O$) and carbon dioxide ($CO_2$), within the upper meter of the surface. A third instrument, the Martian Radiation Experiment (MARIE), provided important insights into the radiation environment at Mars until the instrument was damaged by a large solar flare event on 28 October 2003. As of late 2006, Odyssey's THEMIS and GRS instruments continue to return data.

The two Mars Exploration Rover (MER) missions, named Spirit and Opportunity, were sent to Mars in 2003 to investigate sites which might contain evidence of ancient water and the early martian climate. Spirit was launched on 10 June 2003, with Opportunity following on 7 July. Both rovers utilized the airbag-cushion landing

Figure 1.6 Spirit image of the Gusev Crater plains. This image was taken shortly after Spirit landed. Columbia Hills, ~3km away, can be seen in the distance. (NASA/JPL/Cornell University.)

Figure 1.7 Meridiani Planum from Opportunity. The bedrock exposed in the walls of Eagle Crater can be seen from this image taken by Opportunity shortly after it landed. Landing platform is seen at the bottom of the image. (NASA/JPL/Cornell University.)

system first developed for MPF. Spirit successfully landed at 14.5692°S 175.4729°E on 3 January 2004 inside the 160-km-diameter Gusev impact crater, selected because it was interpreted to be the site of an old lakebed (Squyres *et al.*, 2004a) (Figure 1.6). Opportunity landed on 25 January 2004 in Meridiani Planum (1.9483°S 354.47417°E) (Figure 1.7), a region containing an abundance of the mineral hematite which often forms in a water-rich environment (Squyres *et al.*, 2004b). Although the primary mission for both rovers was 90 days, both are still operating as of the end of 2006. The rovers each carry a Panoramic Camera (Pancam) to survey their surroundings, a Miniature Thermal Emission Spectrometer (Mini-TES) to determine the mineralogy of soil and rocks, a Rock Abrasion Tool (RAT) which allows them to grind past surface coatings and into the unweathered interiors of rocks, a Microscopic Imager (MI) to obtain magnified views of rocks and soils, an Alpha Particle X-ray Spectrometer (APXS) which can be placed against rocks/soil to determine their chemical compositions, and a Mössbauer Spectrometer to determine iron phase and concentrations in the analyzed materials.

The early Mariner and Viking missions provided our first detailed views of Mars. More recent missions (MGS, Pathfinder, Odyssey, and MER) have expanded these initial views and are allowing us to reconstruct the evolutionary history of Mars' surface and atmosphere at a level that was impossible to conceive of with the early missions. The Mars Reconnaissance Orbiter, which launched 12 August 2005 and entered Mars orbit on 10 March 2006, and several future missions, such as the 2007 Phoenix lander and the 2009 Mars Science Laboratory, promise to continue adding new insights into our understanding of the Red Planet.

### 1.2.2 Soviet/Russian Mars missions

The former Soviet Union also was interested in sending spacecraft to Mars from the early days of space exploration. As was often the case during this period of the Space Race between the USSR and USA, the Soviets were the first to send a spacecraft to Mars. After several failures, the Soviets succeeded with the Mars 1 spacecraft, which launched on 1 November 1962 and flew within 195 000 km of the martian surface on 19 June 1963. The spacecraft was designed to photograph the planet, obtain data on the solar wind during transit, and determine if Mars had a magnetic field. The orientation system for the antenna failed in March 1963 so no data were returned during the actual flyby.

A replica of Mars 1 was launched as Zond 2 on 30 November 1964, but one of its solar panels failed and the power output was only half of what was planned. Communications were lost in May 1965, so once again no data were returned during the probe's flyby of Mars on 6 August 1965. Zond 3 missed the Mars launch window in 1964. It was launched on 18 July 1965 on a trajectory towards Mars orbit, even though the planet would not be in the vicinity at the time of "encounter."

Two Soviet launch attempts in 1969 both failed as did the Kosmos 419 launch in 1971. The Soviets had better luck with the Mars 2 and 3 orbiter/lander spacecraft, launched 19 May and 28 May 1971, respectively. Both achieved Mars orbit in late November, a few weeks after Mariner 9 arrived at Mars. Both Mars 2 and 3 were preprogrammed to take pictures of the planet upon their arrival. Unfortunately the global dust storm of 1971 had completely covered the planet by the time the two Soviet spacecraft arrived. Pictures were taken, but they showed only the featureless face of the planet. Other instruments, however, provided more useful data. The infrared radiometer provided the first indication of surface temperatures and thermal inertia, suggesting the planet's surface was covered by dry dust. Temperature observations of the north polar cap were close to the condensation temperature of carbon dioxide, indicating that $CO_2$ is a major component of the ice caps.

Observations of the atmosphere with the Mars 2 and 3 photometers indicated that the altitude of the dust storm clouds was ∼10 km and that the amount of water vapor in the atmosphere during the dust storm was extremely low. Sizes of dust particles (a few microns) and particles in the polar condensation clouds (sub-microns) were made, and radiation scattered from the atmosphere (particularly the Lyman alpha line of hydrogen and the atomic oxygen triplet) was measured. The radio occultation experiment on Mars 2 revealed two major regions of the martian ionosphere, separated by composition and scale height. Measurements of the martian magnetic field indicated a very weak field, with a magnetic moment ∼4000 times smaller than Earth's. Both spacecraft continued operating for four months in orbit, although their orbits provided only seven opportunities for close-up observations. The telemetry of Mars 2 was of very low quality so most of the data from this spacecraft were lost.

Both Mars 2 and 3 also carried landers (called "descent modules" by the Soviets) which were deployed shortly before each orbiter entered Mars orbit. The Mars 2 descent module was released on 27 November but it failed during descent and crashed. The Mars 3 descent module was released on 2 December and successfully descended to the martian surface. After 20 seconds of transmission, however, communication from the lander ceased and was never recovered.

The Mars 4 and 5 orbiters were launched during the 1973 launch window, on 21 July and 25 July, respectively. These were followed by the launches of the Mars 6 and 7 lander carriers on 5 August and 9 August. All four spacecraft arrived at Mars between February and March of 1974. Mars 4 suffered a propellant loss which prevented it from entering Mars orbit. It flew by at 2200 km from the surface, but its radio occultation data detected the nighttime ionosphere for the first time. Mars 5 entered orbit successfully but survived only 22 orbits because of loss of pressurization in the compartment containing the transmitter. The spacecraft returned 60 photographs covering the same area photographed by Mariner 6 in 1969. Mars 5 was the first spacecraft to determine that thermal inertia varies across the surface of Mars, indicating differences in particle size of surface materials. Gamma-ray spectrometry showed that uranium (U), thorium (Th), and potassium (K) concentrations were similar to terrestrial mafic rock concentrations. Data from Mars 4 and 5 revealed that the surface pressure was 670 Pa. Mars 5 also found higher water-vapor levels in the martian atmosphere than had been detected by Mars 3 during the dust storm. Ozone was detected near 40 km altitude, but the concentration was about three orders of magnitude smaller than ozone values on Earth. The US Mariner missions had detected ozone near the martian poles, but Mars 5 also detected it near the equator, although in smaller concentrations. Altitudes and temperatures of the atmospheric layers were better constrained by Mars 5 observations and three plasma zones near the planet were detected.

Neither of the descent modules launched in 1973 survived to the surface. The descent module from Mars 7 missed the planet completely. The Mars 6 descent module sent back information during descent, providing the first *in situ* observations of the atmospheric density and temperature profiles. However, the signal promptly ceased upon landing, suggesting that the lander crashed.

The Soviets did not return to Mars until 1988, focusing their space exploration efforts on Venus in the interim. The Phobos 1 and 2 spacecraft were launched on 7 July and 12 July 1988, respectively, and, as the spacecraft names suggest, the main target was actually Mars' largest moon, Phobos. The spacecraft would eventually achieve a circular orbit around Phobos and send two small stationary landers to the surface to study the elemental composition of the moon, and the orbiters would use laser and ion guns to evaporate small pieces of soil to measure the surface composition using mass spectrometers. Phobos 2 also contained a small lander, called "Hopper," which could hop across the surface of Phobos and measure the elemental composition in different locations.

Phobos 1 received an erroneous command sequence which switched off the communications system less than two months after launch. This problem was never resolved and the spacecraft was lost. Phobos 2 entered Mars orbit on 29 January 1989 with a series of orbit corrections planned over the next two months to bring the spacecraft into orbit around Phobos. Phobos 2 made a number of observations of Mars, including the first thermal infrared wavelength images of the surface by the TERMOSKAN instrument. New information about topography and mineralogy of the martian surface was obtained from the near-infrared mapping spectrometer (ISM) and the thermal infrared radiometer (KRFM) revealed atmospheric features due to aerosol formation such as clouds and limb brightenings. The occultation spectrometer (AUGUST) provided detailed information on the daily variations in atmospheric ozone and water vapor. Although communications were lost just as the Phobos 2 spacecraft was maneuvering into orbit around Phobos, the spacecraft provided detailed images of the moon, the first accurate estimates of the moon's mass and density, information about the surface composition, and temperature variations across the surface. The density and compositional information showed that Phobos was not as similar to carbonaceous chondrites as had been previously believed and that the moon's interior must be more porous than had been expected.

The most recent attempt by the Russians to reach Mars was in 1996 with the Mars 96 mission. The mission contained an orbiter, two surface landers, and two penetrators. It was an ambitious mission, designed to study the topography and mineralogic/elemental composition of the martian surface, study the martian climate and monitor variations in temperature, pressure, aerosols, and composition over time, study the internal structure of the planet, and characterize the plasma environment near Mars. The mission launched 16 November 1996, but a failure of the booster rockets caused the mission to crash into the Pacific Ocean.

### *1.2.3 European Space Agency Mars missions*

The European Space Agency (ESA) has long been interested in sending a mission to Mars and several concept studies have been completed over the years. The first Mars mission for which funding was approved was Mars Express (Chicarro, 2002). Mars Express consisted of an orbiter and the Beagle 2 surface lander. The combined orbiter/lander spacecraft was launched on 2 June 2003 and reached Mars orbit on 25 December 2003. The Beagle 2 lander separated from the orbiter six days before Mars orbit insertion. Beagle 2 was to descend through the atmosphere, slowed by a parachute, and land on the surface using the airbag-type design utilized for Mars Pathfinder and the Mars Exploration Rover missions. It was designed to study the geology and mineralogy of its landing site in Isidis Planitia, study the weather and climate at the landing site, and search for signatures of life. Unfortunately no signal was detected from Beagle 2 after its landing and the lander was declared lost on 6 February 2004.

The Mars Express orbiter, however, has been very successful. All of its instruments have checked out and are returning data. The instruments include several that were lost on the Mars 96 mission, including the High Resolution Stereo Camera (HRSC), the Observatoire pour la Minéralogie, l'Eau, les Glaces, et l'Activité (OMEGA) Visible and Infrared Mineralogical Mapping Spectrometer, the Spectroscopy for Investigation of Characteristics of the Atmosphere of Mars (SPICAM) Ultraviolet and Infrared Atmospheric Spectrometer, the Planetary Fourier Spectrometer (PFS), and the Analyzer of Space Plasma and Energetic Atoms (ASPERA) energetic neutral atoms analyzer. The Mars Radio Science Experiment (MaRS) and Mars Advanced Radar for Subsurface and Ionospheric Sounding (MARSIS) instrument are new experiments on Mars Express.

HRSC has produced spectacular three-dimensional images of the martian surface (Figure 1.8) and is providing important new insights into the surface ages and geologic evolution of the planet. OMEGA has been providing evidence of water ice in the martian polar caps and mineralogic variations across the planet's surface. SPICAM has recently reported nighttime light emissions in the upper atmosphere of Mars due to nitrogen oxide production and the presence of aurorae over the regions retaining strong remnant magnetization (as revealed by MGS's magnetometer experiment). One of the most exciting results from Mars Express has been the detection of atmospheric methane ($CH_4$) by the PFS experiment – the short lifetime of $CH_4$ in the martian atmosphere indicates that this molecule must be replenished by currently active volcanic or hydrothermal processes or perhaps by biologic activity. ASPERA is providing detailed measurements of the plasma environment and upper atmosphere of Mars and has confirmed that the solar wind erodes the upper atmosphere of Mars, providing a mechanism for the loss of water from the

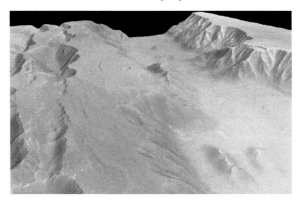

Figure 1.8 Valles Marineris canyon system from HRSC. Stereo images are produced from Mars Express' High Resolution Stereo Camera (HRSC). This stereo image reveals details of Coprates Canyon within the Valles Marineris canyon system. (Image SEMPH01DU8E, ESA/DLR/FU Berlin (G. Neukum).)

planet. MaRS is giving new insights into the structure of the martian atmosphere. MARSIS was deployed in late 2005 and has resolved layering within the polar deposits and buried impact basins. All of these instruments promise to make major changes to our understanding of our neighboring world.

### 1.2.4 Japanese missions to Mars

Japan joined the international exploration of Mars in 1998 with the launch of the Planet B mission. After launch, the mission was renamed Nozomi, Japanese for "hope." The mission was designed to study the upper atmosphere of Mars and its interactions with the solar wind. The mission launched on 3 July 1998 and was originally expected to enter Mars orbit on 11 October 1999. However a malfunctioning valve caused the spacecraft to lose some fuel and two subsequent course-correction burns used more fuel than expected. Since Nozomi no longer had enough fuel for its originally planned trajectory, Japanese engineers put the spacecraft into a heliocentric orbit for four years, allowing it to use two more Earth gravity assists to get it to Mars on a slower trajectory. Its new arrival date at Mars was 14 December 2003. On 9 December 2003, efforts to orient the spacecraft for its orbit insertion five days later failed. Nozomi was put into a heliocentric orbit and all attempts to have it enter Mars orbit were abandoned.

## 1.3 Mars' orbital properties

### 1.3.1 Orbital elements

Mars is the fourth planet from the Sun, with an average distance of $2.279 \times 10^8$ km (1.5237 AU). Mars' orbit is inclined $1.850°$ to the ecliptic plane. Its orbit is among the most elliptical of the planetary orbits, with an eccentricity of 0.0934. This high

Table 1.2 *Orbital properties of Mars*

| | |
|---|---|
| Semimajor axis | $2.2792 \times 10^8$ km |
| | 1.52371034 AU |
| Eccentricity | 0.0933941 |
| Inclination | 1.84969142° |
| Longitude of ascending node | 49.55953891° |
| Longitude of perihelion | 336.0563704° |
| Sidereal orbital period | 686.98 days |
| Synodic orbital period | 779.94 days |
| Mean orbital velocity | 24.13 km s$^{-1}$ |
| Maximum orbital velocity | 26.50 km s$^{-1}$ |
| Minimum orbital velocity | 21.97 km s$^{-1}$ |
| Obliquity | 25.19° |

*Source:* JPL Solar System Dynamics page: ssd.jpl.nasa.gov/ and NSSDC Mars FactSheet: nssdc.gsfc.nasa.gov/planetary/ factsheet/marsfact.html

eccentricity causes a noticeable difference in Mars' perihelion and aphelion distances. The equation of an ellipse in polar coordinates is

$$r = \frac{a(1 - e^2)}{(1 + e \cos \theta)} \tag{1.2}$$

where $a$ is the semimajor axis of the orbit, $e$ is the eccentricity, and $\theta$ is the true anomaly (the angular distance from perihelion to the planet's location, in the direction of orbital motion). Since Mars' average distance from the Sun is its semimajor axis ($a = 2.279 \times 10^8$ km) and we know its orbital eccentricity ($e = 0.0934$), we can calculate the distance of Mars at perihelion ($\theta = 0°$) and aphelion ($\theta = 180°$). At perihelion ($q$), the above equation reduces to

$$q = a(1 - e), \tag{1.3}$$

giving a value of $2.066 \times 10^8$ km (1.381 AU). At aphelion ($Q$), we get

$$Q = a(1 + e). \tag{1.4}$$

Mars' aphelion distance is thus $2.492 \times 10^8$ km (1.666 AU). A summary of Mars' orbital properties for equinox 2000 is provided in Table 1.2.

From Kepler's third law of planetary motion

$$P^2 = \frac{4\pi^2 a^3}{G(M_{Sun} + M_{Mars})} \tag{1.5}$$

we can determine the orbital period ($P$) of Mars from its semimajor axis ($a$) ($G$ is the universal gravitational constant $= 6.67 \times 10^{-11}$ N m$^2$kg$^{-2}$, $M_{Sun}$ is the mass of the

Sun, and $M_{Mars}$ is the mass of Mars). The result is 686.98 Earth days. Mars' mean orbital velocity around the Sun is 24.13 km s$^{-1}$, giving rise to a mean orbital motion of 0.52405° per day. Its maximum orbital velocity, achieved at perihelion, is 26.50 km s$^{-1}$, while its minimum orbital velocity is 21.97 km s$^{-1}$ at aphelion.

Mars' equator is inclined relative to its orbital plane due to the tilt (obliquity) of the rotation axis. Currently the obliquity of Mars is 25.19°. However, as noted in Section 7.4.2, Mars' obliquity (as well as its orbital eccentricity and inclination) vary with time due to gravitational perturbations from the other planets, particularly Jupiter. Currently the position of Mars' north celestial pole is near right ascension $21^h 8^m$, declination $+52°$ 53', close to the star Alpha Cygni (Deneb).

### 1.3.2 Mars' orbital properties with respect to the Sun and Earth

The vernal equinox on Mars is offset about 85° from Earth's vernal equinox. The longitude of the Sun ($L_S$) as seen from Mars ranges from 0° through 360°. For the vernal equinox, the first day of spring in the northern hemisphere, $L_S = 0°$. Northern hemisphere summer solstice occurs at $L_S = 90°$, northern autumnal equinox is at $L_S = 180°$, and northern winter solstice occurs at $L_S = 270°$ (Carr, 1981). Mars' perihelion occurs near $L_S = 250°$, close to summer solstice in the southern hemisphere. Mars' high orbital eccentricity results in the seasons having different lengths. Northern spring (southern autumn) lasts for 199.6 Earth days, northern summer (southern winter) is 181.7 Earth days in length, northern autumn (southern spring) is 145.6 Earth days long, and northern winter (southern summer) lasts 160.1 Earth days.

Mars and Earth reach opposition (when the two planets are aligned on the same side of the Sun) approximately every 779 Earth days. The two planets are closest near this time, although the orbital eccentricities and inclinations can separate the date of closest approach from opposition by a few days. The Earth–Mars distance at opposition varies between $56 \times 10^6$ km and $101 \times 10^6$ km, depending on where Earth and Mars are in their orbits. The closest approaches occur at perihelic oppositions, when Earth is near aphelion and Mars is near perihelion. These happen approximately every 17 years. The furthest distances occur at aphelic oppositions when Earth is near perihelion and Mars near aphelion. A list of all Mars oppositions between 2000 and 2020 is given in Table 1.3. Mars' apparent diameter as seen from the Earth ranges from over 25 arcseconds ($''$) at perihelic oppositions to just under 14$''$ at aphelic oppositions.

## 1.4 Physical properties of Mars

### 1.4.1 Rotation

Mars' physical properties are listed in Table 1.4. Mars' sidereal rotation period is $24^h 37^m 22.65^s$, slightly longer than the terrestrial period of $23^h 56^m 4.09^s$. The

Table 1.3 *Mars oppositions (2000–2020)*

| Opposition date | Closest approach date | Closest distance ($10^6$ km) | Apparent diameter (arcseconds) |
|---|---|---|---|
| 2001 June 13 | June 21 | 67.34 | 20.79 |
| 2003 August 28 | August 27 | 55.76 | 25.11[a] |
| 2005 November 7 | October 30 | 69.42 | 20.19 |
| 2007 December 24 | December 18 | 88.17 | 15.88 |
| 2010 January 29 | January 27 | 99.3 | 14.1 |
| 2012 March 3 | March 5 | 100.8 | 13.9[b] |
| 2014 April 8 | April 14 | 92.4 | 15.2 |
| 2016 May 22 | May 30 | 75.3 | 18.6 |
| 2018 July 27 | July 31 | 57.6 | 24.3 |
| 2020 October 13 | October 6 | 62.2 | 22.6 |

[a] Perihelic opposition.
[b] Aphelic opposition.
*Source:* Students for the Exploration and Development of Space (SEDS) website: www.seds.org/~spider/spider/Mars/marsopps.html

Table 1.4 *Physical properties of Mars*

| | |
|---|---|
| Mass | $6.4185 \times 10^{23}$ kg |
| Volume | $1.6318 \times 10^{11}$ km$^3$ |
| Mean density | 3933 kg m$^{-3}$ |
| Mean radius | 3389.508 km |
| Mean equatorial radius | 3396.200 km |
| North polar radius | 3376.189 km |
| South polar radius | 3382.580 km |
| Sidereal rotation period | 24.622958 h |
| Solar day | 24.659722 h |
| Flattening | 0.00648 |
| Precession rate | $-7576$ milli-arcseconds yr$^{-1}$ |
| Surface gravity | 3.71 m s$^{-2}$ |
| Escape velocity | 5.03 km s$^{-1}$ |
| Bond albedo | 0.250 |
| Visual geometric albedo | 0.150 |
| Blackbody temperature | 210.1 K |

*Source:* Data from Smith *et al.* (2001a) and NSSDC Mars Fact Sheet: nssdc.gsfc.nasa.gov/planetary/factsheet/marsfact.html

direction of rotation is the same as on Earth, counterclockwise as seen from above the ecliptic. The Earth's solar day is defined to be exactly $24^h$; Mars' solar day is $24^h$ $39^m$ $35^s$ and is called a "sol." In the remainder of this book, we will use the term "days" to refer to Earth days and "sols" to refer to Mars days.

Mars' rotation axis undergoes precession due to the gravitational influences of the other planets. From observed changes in the martian rotation rate measured by the Viking landers in the 1970s and the Mars Pathfinder mission in 1997, the precession rate is $-7576\pm35$ milli-arcseconds per year (Folkner *et al.*, 1997). The rate of change of right ascension in the martian equatorial coordinate system due to precession is $-0.1061°$ per century and the rate of change of declination is $-0.0609°$ per century (Folkner *et al.*, 1997).

### *1.4.2 Size*

A detailed discussion of Mars' shape as revealed from geophysical measurements is given in Chapter 3. The current best estimate of the equatorial radius of Mars as measured from the center of figure of the planet is 3397 km (Smith *et al.*, 1999). The radius measured through the poles is slightly less, as is typical for rapidly rotating planets which display an equatorial bulge. Mars' polar radius is 3375 km. The flattening ($f$) of Mars is obtained by comparing its equatorial radius ($R_{equatorial}$) and its polar radius ($R_{polar}$):

$$f = \frac{(R_{equatorial} - R_{polar})}{R_{equatorial}}. \tag{1.6}$$

This gives a value of 0.00648. The mean radius of Mars, which is used in most discussions, is 3390 km.

### *1.4.3 Mass and density*

Mars' mass has been determined from analysis of the orbits of its two moons and the multitude of orbiting spacecraft. The planet has a mass of $6.4185\times10^{23}$ kg, about 11% of the Earth's mass. The shape of Mars will be discussed in Chapter 3, but from geophysical analysis the volume is calculated to be $1.6318\times10^{11}$ km$^3$. That gives a mean density of 3933 kg m$^{-3}$, indicating a rocky body with a small iron core.

## 1.5 Martian moons

Mars' two small moons were discovered during the 1877 perihelic opposition. Asaph Hall of the US Naval Observatory discovered the outermost moon (Deimos) on 12 August 1877, followed by the inner moon (Phobos) six nights later. Hall was becoming frustrated in his attempts to discover any satellites of Mars and was ready to give up. His wife, Chloe Angeline Stickney Hall, convinced him to keep searching a little longer, which led to the moons' discovery. In honor of his wife's persuasion, the largest crater on Phobos is named Stickney (the second largest crater on Phobos is named Hall after Asaph). Hall named the two moons after the

Figure 1.9 Phobos as seen from HRSC. Mars' largest moon, Phobos, displays a heavily cratered and fractured surface. Stickney crater (left) and the fractures radiating from it are visible in this image taken by HRSC on Mars Express. (Image SEMPVS1A90E, ESA/DLR/FU Berlin (G. Neukum).)

mythological sons of the Greek war god Ares: Phobos, the personification of fear or fright, and Deimos, representing dread or terror.

### *1.5.1 Phobos*

Phobos is the larger and innermost of the two moons. It has a very irregular shape (Figure 1.9) with best-fit triaxial ellipsoid radii of $13.3 \times 11.1 \times 9.3 \pm 0.3$ km (Batson *et al.*, 1992). The moon has a mass of $1.06 \times 10^{16}$ kg and a density of 1900 kg m$^{-3}$. This low density suggests that Phobos is either very porous, perhaps with large amounts of ice in its interior, or is a rubble-pile object similar to many asteroids, where the solid pieces are weakly held together by the moon's gravity.

The semimajor axis of Phobos around Mars is 9378.5 km. At this distance it takes Phobos 0.31891 days to orbit Mars – thus, Phobos makes almost three complete orbits in the span of one Martian sol. Its orbit has an eccentricity of 0.01521 and an inclination of 1.08° to Mars' ecliptic plane. Phobos is in synchronous rotation about Mars, so its sidereal period equals its orbital period of 0.31891 days.

Because Phobos orbits inside the geosynchronous point above Mars (i.e., it orbits faster than Mars rotates), tidal interactions between Phobos and Mars are stealing energy from Phobos' orbit. As a result, the moon is slowly spiraling inward towards Mars at a rate of 1.8 meters per century. In about $50 \times 10^6$ years, Phobos will either impact the surface of Mars or be torn apart into a ring system as it passes through Mars' Roche limit.

Phobos is a very dark body, with a geometric albedo of only 0.07. The moon is very heavily cratered, indicating that no internal geologic activity has occurred since its formation. Phobos also displays numerous grooves up to 20 km long across its

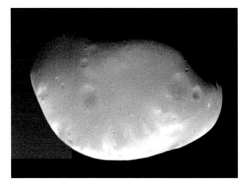

Figure 1.10 Mars' smaller moon, Deimos, is revealed in this mosaic of images from the Viking 1 orbiter. Deimos displays craters, but its surface is subdued because of a thick regolith layer. (NASA/JPL.)

surface, which may be the surface expressions of deeper fractures associated with the porous interior (Thomas *et al.*, 1992a). Many of these grooves are near the crater Stickney, indicating that the formation of this large crater violently shook and stressed the moon, creating new fractures and widening pre-existing ones. Large blocks ejected during crater formation can also be seen on Phobos' surface (Thomas *et al.*, 2000a).

### 1.5.2 *Deimos*

Deimos is the smaller and more distant of the two martian moons. Like Phobos, it is irregular in shape (Figure 1.10), with best-fit triaxial ellipsoid radii of $7.6 \times 6.2 \times 5.4 \pm 0.5$ km (Batson *et al.*, 1992). The mass of Deimos is $2.4 \times 10^{15}$ kg and its density is 1750 kg m$^{-3}$, suggesting another porous or rubble-pile body just like Phobos.

Deimos orbits Mars at a distance of 23 458.8 km, giving it an orbital period of 1.26244 days. The orbit is close to circular, with an eccentricity of 0.0005. The orbital inclination is 1.79°. Deimos is also in synchronous rotation about Mars, with a sidereal rotation period of 1.26244 days.

Deimos is another very dark body, with a geometric albedo of 0.08. It also displays a heavily cratered surface, but the surface is much more muted than that of Phobos where craters and grooves appear relatively sharp. Deimos' subdued appearance likely results from a thick regolith layer covering the surface. This regolith is composed of the fragmented material produced by impact cratering on the moon. The lack of a thick regolith on Phobos may be the result of the violent shaking of the moon during the impact which created Stickney – Phobos' escape velocity is only 0.0057 km s$^{-1}$ so it would not take much energy to launch material into space.

Alternately, Deimos may have different surface properties that preferentially produce large amounts of regolith during crater formation.

### 1.5.3 Origin of Phobos and Deimos

The small, irregular shapes of Phobos and Deimos combined with their very dark surfaces and spectral similarities to C-class asteroids have led many scientists to propose that the two moons are actually asteroids which were captured by Mars. However, dynamic considerations cannot easily reproduce the observed orbits of the two satellites if they are captured asteroids, suggesting instead that the two moons may have formed in the vicinity of Mars.

The asteroidal origin of Phobos and Deimos is supported by the physical and chemical characteristics of the two moons. Both display spectral characteristics similar to those of C-class asteroids. Hartmann (1990) argues that the two moons formed at ~3 AU, within the region dominated by C-class asteroids, and were subsequently scattered inward by gravitational perturbations from Jupiter. The low densities of both moons are similar to those of the most volatile-rich carbonaceous chondrites, which are believed to originate from C-class asteroids. However, if the moons are rubble-pile objects or contain large quantities of water ice in their interiors, their observed densities may not be diagnostic of their compositions.

Spectral reflectance studies of Phobos and Deimos show a very low geometric albedo, similar to the black, neutral objects seen in the outer parts of the main asteroid belt. However, Phobos displays at least four discernible spectral units, one of which is comparable to anhydrous, optically darkened ordinary chondrites (sometimes called black chondrites), which tend to have higher densities than Phobos. Phobos has a heterogeneous surface, with up to 10% variation in albdeo, which results from either compositional variations due to impact cratering processes or differences in grain size distribution (Murchie and Erard, 1996; Simonelli *et al.*, 1998). Deimos is spectrally similar to C- and P-class asteroids.

The major argument against the asteroidal origin of Phobos and Deimos is the difficulty of capturing objects into the near-circular equatorial orbits they display today. Gravitational capture of bodies usually results in orbits with high eccentricities and inclinations. The other problem with the capture model is how to capture the two objects into orbits on opposite sides of the synchronous point. Integrating backwards in time suggests that Phobos could have originated in a much more elliptical orbit at ~5.7 Mars radii ($R_\delta$) – this would be consistent with an origin as a captured asteroid. However, similar integrations for Deimos show it to remain close to its present position and in a near-circular orbit throughout the history of the Solar System, indicating it is most likely not a captured asteroid. In addition, if Phobos' orbit was larger and more eccentric early on, it would have crossed Deimos' orbit, leading to collisions between the two objects. A few models avoid this problem by suggesting that Phobos and Deimos originated as a single body near the synchronous point which was destroyed by an impact event, leaving the two major fragments just inside and just outside the synchronous location (Burns, 1977).

A possible mechanism that would allow Phobos and Deimos to originate as asteroids and eventually be captured into the orbits we see today involves not tidal capture but aerodynamic capture within the early solar nebula (Pollack *et al.*, 1979). In this model, Phobos and Deimos form near ~3 AU in the asteroid belt, are perturbed into Mars-crossing orbits by Jupiter, and then experience aerodynamic drag through interactions with the solar nebula, causing the two bodies to be captured by Mars. Aerodynamic drag causes the orbital semimajor axis, eccentricity, and inclination to decrease, which could produce the observed orbital characteristics of the two moons.

### *1.5.4 Possible other satellites?*

Are Phobos and Deimos the remains of a once larger swarm of small satellites orbiting Mars? Schultz and Lutz-Garihan (1982) suggested that the large number of elliptical craters observed on the martian surface resulted from a swarm of small moons which orbited Mars. The orbits of these moons lay within the synchronous point, causing their orbits to decay over time and resulting in the moons crashing obliquely into the surface to produce the elliptical craters. However, Bottke *et al.* (2000) determined that Mars does not have an anomalously high fraction of elliptical craters compared to the Moon, suggesting that heliocentric, not areocentric (Mars-centered), impactors are primarily responsible for the population of elliptical craters seen on Mars.

### *1.5.5 Trojan asteroids at Mars orbit*

Although Trojan asteroids in Jupiter's L4 and L5 Lagrange points have been known since 1906, Trojan asteroids in terrestrial planet orbits were not discovered until recently. The first Mars Trojan was discovered in 1990 and five more have been discovered as of 2006. Five of the Trojans are located at the L5 point, the other at the L4 point. Spectroscopic analysis of three of the Mars Trojans shows that two are similar to Sr/A asteroids while the third is classified as an X-class asteroid (Rivkin *et al.*, 2003). All three of these asteroids appear to have silicate compositions, based on their spectroscopic classifications, quite different from the carbonaceous compositions of Phobos and Deimos. Numerical models suggest that these Trojan asteroids were captured by Mars early in Solar System history when planetesimals were still abundant (Tabachnik and Evans, 1999). Once the objects are captured into the L4 and L5 points of Mars' orbit, they are dynamically stable over the age of the Solar System (Scholl *et al.*, 2005).

# 2

# Formation of Mars and early planetary evolution

## 2.1 Formation of Mars

### 2.1.1 Accretion

Mars, along with the rest of the Solar System, formed $\sim 4.5 \times 10^9$ years ($10^9$ years $= 1$ Ga) ago out of a cloud of gas and dust called the solar or protoplanetary nebula. The direction of orbital and rotational motions for Mars and most of the other planets and moons indicates that this nebula was slowly rotating in a counterclockwise direction as seen from above the ecliptic plane. The cloud underwent gravitational collapse when it was perturbed by an external event, possibly the explosion of a nearby star, interaction with galactic molecular clouds, or passage through a spiral density wave. Because of the cloud's initial spin motion, the cloud collapsed to a flattened disk with a central bulge. Collapse of the central bulge produced the Sun.

The formation of Mars and the other terrestrial planets is typically divided into three stages: formation of kilometer-sized planetesimals, formation of planetary embryos, and collisional formation of larger planets (Canup and Agnor, 2000; Chambers, 2004). Rotational motion within the flattened disk caused small dust grains ($\sim 1$–30 micrometers [μm] diameter) to collide and stick together, building up larger objects (Weidenschilling and Cuzzi, 1993), although gravitational instabilities causing fragmentation of the cloud into clumps might also have contributed (Ward, 2000; Youdin and Shu, 2002; Chambers, 2004). Gas drag and gravity cause these increasingly larger and more massive particles to settle towards the disk's midplane, explaining why the planetary orbits are approximately coincident with the ecliptic plane. The smaller orbits within the inner Solar System increase the probability of collisions in these areas, allowing inner Solar System bodies to grow faster than objects in the outer Solar System. When these bodies reach a diameter of approximately 1 km, they are called planetesimals.

Planetesimal orbits begin to assume their final characteristics due to a combination of gas drag (causing the orbits to become co-planar and nearly circular) and mutual gravitational interactions (which increase the eccentricity and inclination). The planetesimals have enough mass that their gravity attracts more material to them, a process called accretion. Accretion occurs rapidly, within a time period of $\sim 10^8$ years, and leads to planetesimals growing into planetary embryos (Kokubo and Ida, 1998; Kortenkamp *et al.*, 2000). Planetary embryos have masses similar to that of present-day Mars ($\sim 10^{23}$ kg). Mars was likely prevented from growing larger by the gravitational influence of already-formed Jupiter, which deflected material that could have contributed to Mars' growth into other parts of the Solar System (Chambers and Wetherill, 1998; Chambers, 2001; Lunine *et al.*, 2003). Based on its size and density, Mars probably reached its present size within approximately $10^5$ years (Wetherill and Inaba, 2000).

The subsequent evolution of a planetary embryo into the final planet is largely dictated by late-stage impact events. These impacts can either increase the embryo's size and mass through accretion or decrease the size and mass through erosion, depending on the impact parameters (Canup and Agnor, 2000; Chambers, 2001). Collisions such as that which formed the Earth's moon were common during this period (Hartmann and Davis, 1975; Benz *et al.*, 1989; Cameron, 1997, 2000; Agnor *et al.*, 1999; Kokubo *et al.*, 2000). The planet's final characteristics, such as mass and rotation rate, are largely determined by this late accretionary phase (Agnor *et al.*, 1999; Lissauer *et al.*, 2000; Canup and Agnor, 2000). Mars' high concentration of volatiles (Section 2.3) and small size suggest that it was not strongly affected by this large impact stage (Chambers, 2004).

Temperature and pressure varied throughout the solar nebula, resulting in different elements condensing in different locations. Equilibrium condensation diagrams show that the decrease in temperature outward from the Sun causes elements which cannot be easily vaporized (refractory elements) to condense near the Sun while easily vaporized elements (volatile elements) condense at greater distances. Iron oxides should be present in the solar nebula near Mars' distance while water ice condenses closer to Jupiter. The equilibrium condensation sequence provides insights into what elements should have been present in the material which formed each of the planets.

Although the "ice line" for water ice occurs close to Jupiter, hydrated minerals can form closer to the Sun (Drake and Righter, 2002). These hydrated minerals have $H_2O$ or hydroxyl (OH) molecules attached to their basic unit. Although many scientists believe that much of the water seen in the inner Solar System was provided by asteroid and comet impacts (Morbidelli *et al.*, 2000), some hydrated minerals could have been incorporated into the growing terrestrial planets directly from the solar nebula.

## *2.1.2 Heavy bombardment*

Not all of the material present in the solar nebula was immediately accreted into the planets. A significant amount of material still remained after the planets had formed. This material underwent gravitational and collisional perturbations which caused it to cross the orbital paths of the newly formed planets. Occasionally this debris from planetary formation collided with the newly formed planet, creating impact craters on bodies with solid surfaces. The impact rates during this time are estimated to be between 100 and 500 times what they are today and this period of higher impact cratering rates is therefore called the Period of Heavy Bombardment.

The Period of Heavy Bombardment is divided into an early and a late period. The Early Heavy Bombardment Period includes the period from planet formation until the surface solidifies. No cratered surfaces remain from this time period. The Late Heavy Bombardment (LHB) Period is the period from when the surface solidified until the impact rate declined to the current level. Scars of the LHB are still seen on the heavily cratered surfaces of the Moon, Mercury, and Mars (Figure 2.1). Based on analysis of lunar samples returned by the Apollo and Luna missions, the end of the LHB is estimated to have occurred about 3.8 Ga ago in the Earth–Moon system and probably close to this time elsewhere in the inner Solar System (Section 5.1.3).

Traditionally the LHB has been envisioned as an exponential decline in impact rates over the period between ~4.5 and ~3.8 Ga ago (Figure 2.2). However, analysis of lunar highland rocks, lunar impact melt rocks, and lunar meteorites found that none of these samples displayed an age older than ~4.0 Ga, suggesting that the LHB was actually a short, intense period of bombardment (Tera *et al.*, 1974; Dalrymple and Ryder, 1993, 1996; Cohen *et al.*, 2000; Culler *et al.*, 2000; Stöffler and Ryder, 2001; Kring and Cohen, 2002). This idea of a lunar cataclysm around 3.9 Ga ago has recently received additional support from dynamical modeling, which suggests that the LHB in the inner Solar System was triggered when an outer planetesimal disk was destabilized during migration of the giant planets ~0.7 Ga after the Solar System formed (Gomes *et al.*, 2005). Others, however, argue that the lack of lunar samples with ages >4.0 Ga simply results from destruction of such material by later impacts and that the sampled material is largely derived from the Imbrium impact event (Hartmann, 2003). Whether comets or asteroids were the dominant impactors during the LHB is unclear (Gomes *et al.*, 2005), although lunar sample geochemical analysis (Kring and Cohen, 2002) and comparison of the crater size–frequency distribution curves with the size–frequency distribution of main belt asteroids (Strom *et al.*, 2005) suggests that asteroids may have dominated the LHB populations of impacting bodies in the inner Solar System.

(a)

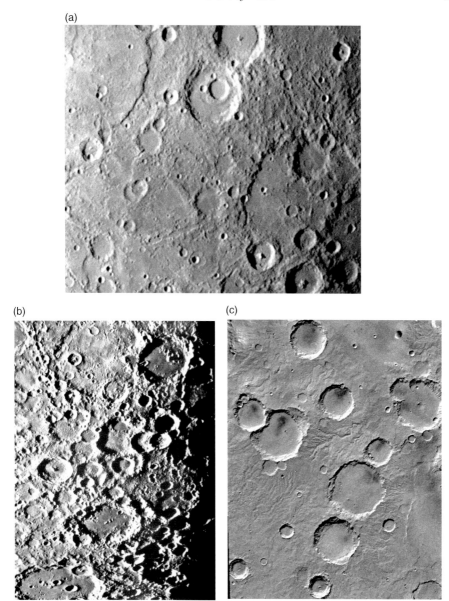

(b)

(c)

Figure 2.1 Heavily cratered surfaces are seen on Mercury, the Moon, and Mars in the inner Solar System. These regions display high crater densities, indicating they retain the scars of impacts occurring during the Late Heavy Bombardment Period. (a) Mariner 10 image of Mercury near the planet's south pole. (Image PIA02937; NASA/JPL/Northwestern University.) (b) This image of the region near the Moon's south pole was obtained from Lunar Orbiter images. (Image C2435 from the Consolidated Lunar Atlas [Kuiper *et al.*, 1967], courtesy of Lunar and Planetary Institute.) (c) This mosaic from Viking Orbiter images shows a heavily cratered region of Mars centered near 37.5°S 148.5°E. (NASA/JPL.)

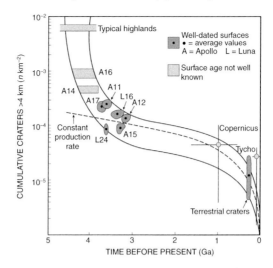

Figure 2.2 The lunar crater chronology curve compares the cumulative crater density to age of the surface as derived from Apollo sample analysis. It allows one to estimate the age of a surface based solely on the crater density. (Reprinted by permission from Cambridge University Press, Heiken *et al.* [1991], Copyright 1991.)

## 2.2 Differentiation and core formation

### 2.2.1 Planetary heating

Early modeling of planetary formation assumed that planets formed cold and heated up through later decay of radioactive elements (Toksöz and Hsui, 1978; Solomon, 1979). However, the realization that accretionary processes dominate planetary formation led astronomers to realize that planets are formed hot and gradually cool over time (Grossman, 1972; Benz and Cameron, 1990; Boss, 1990). The heat comes from two main sources: the kinetic energy of impacting bodies is transformed in part into heat during accretion (accretionary heating), and radioactive elements, particularly short-lived radioisotopes, generate heat during decay. Complete melting of a Mars-sized object requires approximately $2 \times 10^{30}$ joules (J) of energy. For Mars, accretionary heating likely produced $\sim 4 \times 10^{30}$ J (e.g., Wetherill, 1990). Among the short-lived radionuclides, $^{26}$Al likely contributed the most heat, with estimates around $2 \times 10^{30}$ J (Elkins-Tanton *et al.*, 2005). Thus, accretional heating and short-lived radionuclide decay can easily produce enough energy to melt Mars.

This semi-fluid state of the planet allows materials to segregate into layers, a process called differentiation. Differentiation occurs because of density variations and differing chemical affinities of the elements. The variation in density among different elements causes the denser material to sink towards the planet's center while less dense materials rise to form the surface. In particular, iron tends to sink towards the center to form a core while oxygen compounds rise towards the surface to create the crust. Certain elements, called siderophile elements, have a chemical affinity for iron and will follow iron (Fe) to the core. Examples of siderophile elements include nickel (Ni), cobalt (Co), iridium (Ir), and platinum (Pt). Elements that follow oxygen (O) are called lithophile elements; examples include potassium (K), sodium (Na), calcium (Ca), magnesium (Mg), and aluminum (Al). Elements that tend to form compounds with sulfur (S) are called chalcophile elements and include zinc (Zn), copper (Cu), and lead (Pb). The combination of density and chemical affinities explains why iron and sulfur compounds are typically found deep in planetary interiors while planetary crusts are largely composed of oxides.

Information about Mars' early evolution comes from analysis of radioactive elements and their decay products in martian rocks. None of the spacecraft missions to Mars has yet returned soil or rock samples which can be analyzed in terrestrial laboratories. The lander/rover missions are beginning to supply mineralogic information which can provide some insights into the planet's bulk composition, but instrument capabilities limit the amount of information obtained by these missions. However, nature has provided scientists with samples of the martian surface in the form of martian meteorites. Geochemical analysis of the 37 martian meteorites currently in our collections provides important insights into the early history of the martian interior. The thermal evolution of Mars is constrained through both geochemical analysis and geophysical modeling.

### 2.2.2 Geochronology

Many elements have isotopes that are unstable, decaying to a stable isotope in a well-defined period of time. The radioactive element is called the parent element while the stable decay product is called the daughter element. The amount of daughter element at any point in time depends on the original amount of the daughter element, the original amount of the parent element, the rate at which the parent decays into the daughter element (given by the decay constant, $\lambda$), and the amount of time that has elapsed. If $N$ is the number of parent atoms in the sample, then the change in the number of those atoms over time ($dN/dt$) is given by

$$\frac{dN}{dt} = -N\lambda. \tag{2.1}$$

Integrating this gives

$$N_t = N_0 e^{-\lambda(t-t_0)} \tag{2.2}$$

in which $N_0$ is the number of parent atoms at $t = t_0$ and $N_t$ is the number of parent atoms remaining after time $(t - t_0)$. $N_t$ is related to $N_0$ and the amount of daughter element ($D_r$) which has been produced by the decay:

$$N_t = N_0 - D_r. \tag{2.3}$$

If the entire daughter element present in the sample is from the decay of the parent, then $N_t$ and $D_r$ will be the measurable amounts of the parent and daughter in the sample, respectively. The value of $N_0$, the original amount of the parent, cannot be directly determined from the sample analysis, so we substitute Eq. (2.3) into Eq. (2.2) to eliminate $N_0$:

$$N_t = (N_t + D_r)e^{-\lambda(t-t_0)}. \tag{2.4}$$

We can rewrite Eq. (2.4) as a ratio of the daughter and parent concentrations at time $t$:

$$\frac{D_r}{N_t} = e^{\lambda(t-t_0)} - 1. \tag{2.5}$$

The amount of time for half of the original amount of the parent radioisotope to decay into the daughter is called the half-life ($t_{1/2}$). It is obtained by setting $N_t = 1/2\, N_0$ in Eq. (2.2), which then reduces to

$$-\lambda t_{1/2} = \ln(1/2). \tag{2.6}$$

Thus, the decay constant, $\lambda$, and the half-life, $t_{1/2}$, are related to each other by

$$\lambda = \frac{0.693}{t_{1/2}}. \tag{2.7}$$

Igneous rocks typically contain small amounts of radioactive elements and are therefore the most useful rocks for geochronologic purposes. Some of the common radioisotope systems used in geologic analysis are given in Table 2.1.

Equation (2.5) can provide the age of an individual sample for which the concentrations of the daughter and parent elements are measured. However, the uncertainties in ages based on a single measurement are typically quite large. Geochemists therefore determine a rock's age from measurements of several minerals within the sample. The resulting age is the crystallization age of the sample, which indicates the amount of time that has elapsed since the rock solidified.

The amount of the daughter element at some time $t$ ($D_t$) is the sum of the original amount of daughter at $t = t_0$ ($D_0$) and the amount which has been produced by the decay of the parent ($D_r$):

$$D_t = D_0 + D_r. \tag{2.8}$$

Thus, we also need to know the initial (non-radiogenic) amount of daughter to uniquely determine the crystallization age of the sample. While absolute

Table 2.1 *Geologically important radionuclides*

| | Parent | Daughter | Half-life (yrs) |
|---|---|---|---|
| Long-lived radionuclides | $^{147}Sm$ | $^{143}Nd$ | $106\times10^9$ |
| | $^{87}Rb$ | $^{87}Sr$ | $48.8\times10^9$ |
| | $^{187}Re$ | $^{187}Os$ | $46\times10^9$ |
| | $^{232}Th$ | $^{208}Pb$ | $14\times10^9$ |
| | $^{238}U$ | $^{206}Pb$ | $4.47\times10^9$ |
| | $^{40}K$ | $^{40}Ar$ | $1.25\times10^9$ |
| | $^{235}U$ | $^{207}Pb$ | $7.04\times10^8$ |
| Short-lived radionuclides | $^{146}Sm$ | $^{142}Nd$ | $103\times10^6$ |
| | $^{244}Pu$ | $^{240}U$ | $82\times10^6$ |
| | $^{129}I$ | $^{129}Xe$ | $17\times10^6$ |
| | $^{182}Hf$ | $^{182}W$ | $9\times10^6$ |
| | $^{53}Mn$ | $^{53}Cr$ | $3.6\times10^6$ |
| | $^{26}Al$ | $^{26}Mg$ | $7.2\times10^5$ |

concentrations are impossible to determine, the ratio of the daughter isotope to another isotope of the same element which is not produced by radioactive decay is typically a constant for material formed from a particular region of the solar nebula. If we represent this stable isotope of the daughter element by $X$, the left side of Eq. (2.5) becomes

$$\frac{\left(\frac{D_t}{X}\right) - \left(\frac{D_0}{X}\right)}{(N_t/X)}.$$

The ratio $D_0/X$ is constant for all mineral samples and will drop out when comparing the concentrations of $D_t/X$ and $N_t/X$ among the different samples. Therefore, crystallization ages are determined using these ratios rather than absolute concentrations of parent and daughter elements.

   Crystallization ages are determined using isochron diagrams, such as that shown for the rubidium (Rb)–strontium (Sr) system in Figure 2.3. Strontium-86 ($^{86}Sr$) is the non-radiogenic isotope of strontium to which the $^{87}Rb$ and radiogenic $^{87}Sr$ are compared. Assume that at time $t_0$ we have magma solidification occurring. Three minerals in this melt ($A$, $B$, and $C$) have specific $^{87}Rb/^{86}Sr$ concentrations, as shown. While the reservoir is molten, any $^{87}Sr$ produced by the decay of $^{87}Rb$ is free to migrate through the reservoir. Hence the concentration of $^{87}Sr$ in the magma reservoir will be homogeneous. At time $t_0$ when the magma reservoir solidifies, the concentration of $^{87}Sr/^{86}Sr$ is therefore constant, as shown in Figure 2.3. Once the reservoir solidifies, the concentration of $^{87}Sr/^{86}Sr$ increases and that of $^{87}Rb/^{86}Sr$ decreases within each specific mineral due to the decay of $^{87}Rb$. The concentrations of these elements for our three minerals changes as shown by the arrows in Figure 2.3.

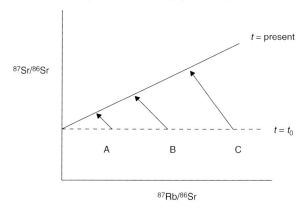

Figure 2.3 Isochron diagram, showing how rubidium (Rb) decays into strontium (Sr) in three minerals (*A*, *B*, and *C*) over a time period $t_0$ to $t$.

At time $t$, we can measure concentrations of $^{87}Rb/^{86}Sr$ and $^{87}Sr/^{86}Sr$ and compare them among the three minerals. From the graph and Eq. (2.5), the slope of the isochron ("equal time") line at time $t$ is given by

$$\frac{\Delta(^{87}Sr/^{86}Sr)}{\Delta(^{87}Rb/^{86}Sr)} = e^{\lambda(t-t_0)} - 1 \qquad (2.9)$$

where $\Delta$ indicates the difference in concentrations between two minerals. This equation gives the crystallization age $(t - t_0)$ of the sample from which the minerals were obtained.

The above technique only provides accurate crystallization ages if the system has been "closed" since the reservoir solidified. A closed chemical system is one where there has been no addition or subtraction of parent and/or daughter elements. Open chemical systems are often seen when either the parent or daughter element is very volatile – an increase in temperature can cause the volatile component to readily escape from the system. To determine whether the system has been open or closed, geochemists look at the whole rock age. The isochron graph for a whole rock age uses the average $^{87}Rb/^{86}Sr$–$^{87}Sr/^{86}Sr$ value for the entire rock. The initial value of $^{87}Sr/^{86}Sr$ is the second point on the graph, allowing the isochron line to be drawn between these two points. The initial $^{87}Sr/^{86}Sr$ value is determined from the oldest age measurements of Solar System material. This measurement comes from basaltic achondrite meteorites, which formed 4.5 Ga ago when the Solar System formed. The $^{87}Sr/^{86}Sr$ value from the basaltic achondrites is called the Basaltic Achondrite Best Initial (BABI) value and is equal to $0.698\,99 \pm 0.000\,04$ (Birck and Allègre, 1978). The model age which results from the whole rock analysis fixes the time of origin of the initial materials comprising the rock, without regard to the rock's subsequent history. If the model age equals the age of the Solar System (4.5 Ga), the system is a

Figure 2.4 EETA 79001 is one of the shergottite martian meteorites. It was the first martian meteorite in which gas from the martian atmosphere was identified. (NASA/Johnson Space Center [JSC].)

closed system. If the model age is not equal to the age of the Solar System, the system has been an open system during at least some part of its lifetime. A model age less than the age of the Solar System indicates that the parent-to-daughter ratio (*N/D*) has increased while an age greater than the age of the solar system indicates that *N/D* has decreased.

### 2.2.3 Martian meteorites

Most meteorites have crystallization ages of 4.5 Ga and are believed to represent material which formed directly from the solar nebula. By 1979, however, three meteorites were known with younger formation ages and distinct mineralogic properties. These three meteorites were seen to fall in 1865 near Shergahti, India, in 1911 near El-Nakhla, Egypt, and in 1815 near Chassigny, France. These three meteorites are now called Shergotty, Nakhla, and Chassigny, respectively, and are the representative samples of the meteorites called the shergottites, nakhlites, and chassignites (SNC) (Figure 2.4).

By 1979, scientists began to speculate that the SNC meteorites are pieces of Mars ejected off the planet during meteorite impacts. All of the SNC meteorites are volcanic rocks and all but one have formation ages <1.35 Ga. The young ages and volcanic textures indicate that these rocks come from a body where volcanism has been occurring up to at least 0.16 Ga ago (age of youngest SNC). Analysis of the oxygen isotopes in these rocks revealed that they are distinct from terrestrial, lunar, and most meteoritic rocks (Clayton and Mayeda, 1996). By 1980, the circumstantial evidence strongly suggested that Mars is the parent body of the SNC meteorites. The 1982 discovery that the isotopic composition of trapped noble gases within the SNC

Table 2.2 *Martian meteorites*

| Meteorite name | Year found | Type | Crystallization age (yrs)[a] |
|---|---|---|---|
| Chassigny | 1815 | Chassignite | $1.3\times10^9$ |
| Shergotty | 1865 | Basaltic shergottite | $165\times10^6$ |
| Nakhla | 1911 | Nakhlite | $1.3\times10^9$ |
| Lafayette | 1931 | Nakhlite | $1.3\times10^9$ |
| Governador Valadares | 1958 | Nakhlite | $1.3\times10^9$ |
| Zagami | 1962 | Basaltic shergottite | $180\times10^6$ |
| ALHA 77005 | 1977 | Lherzolitic shergottite | $185\times10^6$ |
| Yamato 793605 | 1979 | Lherzolitic shergottite | $170\times10^6$ |
| EETA 79001 | 1980 | Olivine–phyric and basaltic shergottite | $180\times10^6$ |
| ALH 84001 | 1984 | Orthopyroxenite | $4.1\times10^9$ |
| LEW 88516 | 1988 | Lherzolitic shergottite | $180\times10^6$ |
| QUE 94201 | 1994 | Basaltic shergottite | $320\times10^6$ |
| Dar al Gani 476, 489, 670, 735, 876, 975[b] | 1997–1999 | Olivine–orthopyroxene shergottite | $474\times10^6$ |
| Yamato 980459 | 1998 | Olivine–phyric shergottite | $472\times10^6$ |
| Los Angeles 001, 002 | 1999 | Basaltic shergottite | $165\times10^6$ |
| Sayh al Uhaymir 005, 008, 051, 094, 060, 090, 120, 150[b] | 1999–2004 | Olivine–phyric shergottite | N.d. |
| Dhofar 019 | 2000 | Olivine–phyric shergottite | $525\times10^6$ |
| GRV 99027 | 2000 | Lherzolitic shergottite | N.d. |
| Dhofar 378 | 2000 | Basaltic shergottite | N.d. |
| Northwest Africa 480, 1460[b] | 2000–2001 | Basaltic shergottite | $340\times10^6$ |
| Yamato 000593, 000749, 000802[b] | 2000 | Nakhlite | $1.3\times10^9$ |
| Northwest Africa 817 | 2000 | Nakhlite | $1.35\times10^9$ |

| | | | |
|---|---|---|---|
| Northwest Africa 1669 | 2001 | Basaltic shergottite | N.d. |
| Northwest Africa 1950 | 2001 | Lherzolitic shergottite | N.d. |
| Northwest Africa 856 | 2001 | Basaltic shergottite | $186 \times 10^6$ |
| Northwest Africa 1068, 1110, 1183, 1775[b] | 2001–2004 | Olivine–phyric shergottite | $185 \times 10^6$ |
| Northwest Africa 998 | 2001 | Nakhlite | $1.3 \times 10^9$ |
| Northwest Africa 1195 | 2002 | Olivine–orthopyroxene shergottite | N.d. |
| Northwest Africa 2046 | 2003 | Olivine–orthopyroxene shergottite | N.d. |
| MIL 03346 | 2003 | Nakhlite | $1.02 \times 10^9$ |
| Northwest Africa 2737 | 2004 | Chassignite | $1.3 \times 10^9$ |
| Northwest Africa 3171 | 2004 | Basaltic shergottite | N.d. |
| Northwest Africa 2626 | 2004 | Olivine–orthopyroxene shergottite | N.d. |
| YA1075 | ?? | Lherzolitic shergottite | N.d. |
| Northwest Africa 2975 | 2005 | Basaltic shergottite | N.d. |
| GRV 020090 | 2005 | Lherzolitic shergottite | N.d. |
| Northwest Africa 2646 | 2005 | Lherzolitic shergottite | N.d. |

[a] N.d., no data available.
[b] Fragmented meteorites.
*Source*: Data from Mars Meteorite Compendium: curator.jsc.nasa.gov/antmet/mcc/index.cfm

meteorites was statistically identical to the martian atmosphere clinched the case for a martian origin of the SNC meteorites (Bogard and Johnson, 1983).

As of 2006, 37 martian meteorites are known (Table 2.2). Of these meteorites 27 are classified as shergottites, seven are nakhlites, and two are chassignites. The last meteorite is ALH84001, an orthopyroxenite which is an apparent sample of the ~4-Ga-old cumulate crust of Mars.

Shergottites are subdivided into four classes, based on composition: basaltic, lherzolitic, olivine–orthopyroxene, and olivine–phyric. Nakhlites are rich in olivine while chassignites are dunites. ALH84001 is the only orthopyroxenite among the known martian meteorites. These 37 meteorites have been shocked to various degrees (pressure range from 30 to 50 GPa) and have apparently been ejected off Mars in five to eight different impact events (Nyquist *et al.*, 2001). The timing of these ejection events can be estimated by the cosmic ray exposure ages of the meteorites – all are $<16 \times 10^6$ years ($10^6$ years $= 1$ Ma). Many different craters have been proposed as the sources of these meteorites, based on geologic analysis (Wood and Ashwal, 1981; Mouginis-Mark *et al.*, 1992a; Barlow, 1997; Tornabene *et al.*, 2006). Comparison of meteorite mineralogies with information from MGS TES and Odyssey THEMIS instruments has been difficult since much of the surface is covered by dust which masks the underlying mineralogy, but a few intriguing areas have been identified by this technique (Hamilton *et al.*, 2003; Harvey and Hamilton, 2005).

One of the perplexing issues related to martian meteorites is why 97% of the meteorites come from geologically young terrains when such surfaces are believed to constitute <40% of Mars' surface area. This martian meteorite paradox was of particular concern when it was thought that all martian meteorites were ejected from a single large impact crater (Melosh, 1984) – such large impact craters are rare on younger surfaces. Although multiple impact sites are now proposed (Treiman, 1995; Nyquist *et al.*, 2001) and smaller impacts are capable of launching the martian meteorite material (J.N. Head *et al.*, 2002), the primary difference in surface ejection capability appears to be that the thicker regolith cover on older surfaces likely inhibits ejection of the martian meteorites from those regions (Hartmann and Barlow, 2006).

### 2.2.4 *Differentiation and core formation on Mars*

Martian meteorite analysis provides important constraints on the early evolution of Mars, particularly in regard to the planet's differentiation and core formation. Most of the results are based on analysis of the shergottites, which, due to their volcanic textures, provide information about mantle conditions.

Based on martian meteorite analysis, Wänke (1981) proposed that Mars accreted from two major reservoirs of material, both with chondritic (C1) abundance ratios. Component A consisted of highly reduced, volatile-poor materials while

Component B was oxidized and volatile-rich. Dreibus and Wänke (1985) argued for homogeneous accretion of these two components in a ratio of 60% Component A to 40% Component B. Planetary formation models suggest that accretion of Mars occurred rapidly, within ~0.1 Ma (Wetherill and Inaba, 2000), and that Mars was largely spared from late-stage large impact events (Chambers and Wetherill, 1998) which affected the interior geochemistry of other bodies.

The addition of heat from accretion, large impacts, and short-lived radioactivity melted the planet and caused differentiation. Evidence for an early differentiation of Mars comes from analysis of $^{182}$W, the daughter product resulting from decay of $^{182}$Hf (Hf: hafnium). Tungsten (W) behaved as a siderophile element early in martian history, following iron (Fe) to the core during core formation. Any $^{182}$W produced after core formation will remain in the mantle. Since the half-life of $^{182}$Hf is 9 Ma, the detection of radiogenic $^{182}$W in mantle materials provides constraints on the timing of differentiation and core formation. The Hf/W ratio is approximately five times lower on Mars than in terrestrial materials, but some radiogenic $^{182}$W has been detected. The W analysis suggests that accretion and differentiation on Mars was complete within 10 to 15 Ma after Solar System formation (Lee and Halliday, 1997; Kleine *et al.*, 2002). This is supported by a correlative relationship in the behavior of neodymium (Nd) and W, suggesting that both underwent fractionation from the mantle material within the first 15 Ma of Solar System history (Halliday *et al.*, 2001). Early core formation is likewise indicated by the observation that the lead (Pb) isotopes ($^{206}$Pb and $^{207}$Pb) follow the U/Pb geochron at 4.5 Ga ago (Chen and Wasserburg, 1986), and that Re/Os analysis suggests early fractionation of rhenium (Re) from osmium (Os) (Brandon *et al.*, 2000).

## 2.3 Bulk composition of Mars

The bulk composition of Mars can be estimated from its density (3933 kg m$^{-3}$) and surface composition. More detailed determination of the bulk composition requires analysis of martian rocks. Although analysis of crustal materials by surface missions (Viking, Mars Pathfinder, and MER) and orbital spacecraft (MGS TES, Odyssey THEMIS and GRS, and Mars Express OMEGA) provides some information, most of our current understanding of the bulk composition of Mars comes from the martian meteorites.

Analysis of martian meteorites, particularly the basaltic shergottites, provides information on both major and minor elemental abundances on Mars, which gives insights into the bulk composition of the planet and how it has evolved over time (Wänke, 1981; Dreibus and Wänke, 1985, 1987; Lodders, 1998). Comparisons between the martian meteorites and terrestrial rocks suggest that while the overall evolution of the two planets shows similarities, there are several important differences. Some of these differences result from the smaller size and higher volatile concentration of Mars (Bertka and Fei, 1997) while others relate to the lack of active plate tectonics throughout most if not all of martian history (Breuer *et al.*, 1997).

Martian meteorite analysis indicates that the planet's mantle is approximately twice as rich in iron as Earth's mantle (Longhi and Pan, 1989; Halliday *et al.*, 2001). Iron in the mantle is expected to exist primarily as FeO, while in the core it is expected to occur in metallic form and to interact with sulfur to form FeS. The absolute concentration of FeO in the martian mantle is ~17.9 (0.6 wt%, compared to the terrestrial value of ~8 wt% [Halliday *et al.*, 2001]). The higher Fe content on Mars likely resulted from more oxidizing conditions during core formation. Mars also exhibits higher concentrations of moderately volatile elements than the Earth; for example, Rb/Sr and K/U concentrations are approximately twice as high on Mars (Longhi *et al.*, 1992). The higher volatile concentrations and more oxidizing conditions on Mars may have resulted from the lack of large impacts during the final stages of accretion, allowing Mars to retain its accreted volatiles.

The higher FeO and volatile concentrations in the martian mantle help to explain many of the unusual isotopic behaviors observed in the martian meteorites. For example, moderately siderophile elements on Earth such as phosphorus (P) become more lithophilic under the oxidizing conditions found on Mars and thus typically show less depletion in martian meteorites than in terrestrial rocks. Because P is not depleted in the martian mantle, phosphates dictate much of the behavior of large ion lithophilic (LIL) elements such as the rare earth elements (REE), uranium (U), thorium (Th), and samarium (Sm), concentrating such elements within phosphate deposits. Analysis of the soil and rock compositions from the Opportunity landing site (Squyres *et al.*, 2004b) and remote sensing of surface mineralogy from OMEGA (Bibring *et al.*, 2006) suggest that acid weathering may have been common on Mars at least periodically throughout its history. Phosphates easily dissolve in slightly acidic solutions, allowing the LIL elements to become mobile and affecting radiometric age dating that utilizes these elements. The presence of phosphates may also help explain the fractionations of Zr/Nb, Lu/Hf, and Sm/Nd observed in the martian meteorites (Blichert-Toft *et al.*, 1999).

Oxygen isotopes are distinctly different between the martian meteorites and rocks from elsewhere in the Solar System. The three major isotopes of oxygen used in these analyses are $^{16}O$, $^{17}O$, and $^{18}O$. The ratio $^{17}O/^{16}O$ is normalized to standard mean ocean water (SMOW) on Earth and is designated by $\delta^{17}O$:

$$\delta^{17}O \equiv \left[ \frac{(^{17}O/^{16}O)_{rock}}{(^{17}O/^{16}O)_{SMOW}} - 1 \right] \times 1000. \qquad (2.10)$$

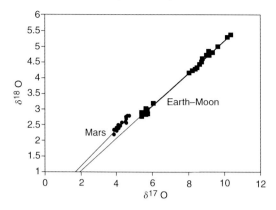

Figure 2.5 Generalized oxygen isotope plot for terrestrial rocks, lunar samples, and martian meteorites. The martian meteorites have a distinct oxygen isotopic pattern, indicating they originated from a region of the Solar System that was more volatile-rich than that found near 1 AU. (Data from Clayton and Mayeda, 1996.)

Similarly, $^{18}O/^{16}O$ is designated by $\delta^{18}O$. A plot of $\delta^{18}O$ vs. $\delta^{17}O$ reveals distinct differences among rocks from Earth, Moon, Mars, and various meteorites (Clayton, 1993) (Figure 2.5). These results represent differences in oxygen mass fractionation within the solar nebula. Thus, the oxygen isotopes provide information about where specific material formed within the Solar System. The coincidence of oxygen isotopes for terrestrial and lunar rocks is one of the indications that the Moon formed primarily of terrestrial material ejected in a large impact event. Oxygen data from the martian meteorites lie above and parallel to the Earth–Moon line and suggest that the planetesimals which formed Mars were more volatile-rich than those near 1 AU.

The SNC meteorites also provide insights into the water content of the martian mantle. Comparing the abundance of water to that of other volatile species and assuming the two component accretionary model of Mars, Dreibus and Wänke (1987) found that the martian mantle contains only 36 parts per million (ppm) of water. Accretionary models suggest that Mars was probably more water-rich when it first formed. The excess water would be removed by reactions with metallic iron early in martian history (Fe + $H_2O \rightarrow$ FeO + $H_2$), creating the oxidized (FeO-rich) mantle and releasing hydrogen, which would dissolve in the core (Zharkov, 1996) or be lost through outgassing and hydrodynamic escape from the atmosphere (Hunten *et al.*, 1987). The dry mantle model of Dreibus and Wänke (1987) initially appears inconsistent with the geologic evidence of substantial amounts of surface water in the martian past (Chapters 5 and 7), but can be explained through a combination of a late-stage volatile-rich veneer emplaced on the planet's surface and lack of plate tectonics that would prevent recycling of surface water into the interior (Carr and Wänke, 1992). Higher water concentrations have been proposed from direct analysis of water contained in the SNCs (Karlsson *et al.*, 1992), analysis of melt inclusions in the SNCs (McSween and Harvey, 1993), and geophysical modeling of a martian magma ocean (Elkins-Tanton and Parmentier, 2006), although the uncertainties associated with these techniques are quite large.

Martian meteorite studies together with results from the Viking, Mars Pathfinder, and MER surface missions indicate that martian rocks contain large amounts of sulfur (S) (Longhi *et al.*, 1992; McSween *et al.*, 1999; Gellert *et al.*, 2004; Klingelhöfer *et al.*, 2004; Rieder *et al.*, 2004; Ming *et al.*, 2006). Sulfur solubility in a silicate melt depends on the iron concentration of that melt (Wallace and Carmichael, 1992). With the higher FeO content of the martian mantle, it is not unexpected that the mantle would also contain large quantities of sulfur. The oxidizing and S-rich conditions within the martian interior cause moderately siderophile elements to display lithophilic behavior, concentrating such elements in the crust and mantle rather than the core. Core formation should have drawn siderophile elements away from the mantle and crust, but martian meteorites show a lower depletion of P, manganese (Mn), chromium (Cr), and W compared to terrestrial rocks, indicating that these elements displayed lithophilic characteristics because of the oxidized and S-rich mantle conditions (Wänke, 1981; Halliday *et al.*, 2001). Only the highly siderophile elements show strong depletions consistent with core formation.

## 2.4 Thermal evolution of Mars

Early thermal models suggested that Mars accreted cold and took up to 2 Ga to complete differentiation and core formation (Johnston *et al.*, 1974; Toksöz and Hsui, 1978). Improved understanding of the dynamics of accretionary processes have led to the realization that Mars formed hot due to rapid accretion (within ~5 Ma [Wetherill and Inaba, 2000]), with differentiation and core formation completed within ~20 Ma after Solar System formation (Lee and Halliday, 1997). Accretional heating, core formation, and decay of radioactive elements produced sufficient energy to melt Mars, leading to a magma ocean (Elkins-Tanton *et al.*, 2005). The combination of geochemical information about the bulk composition of Mars, geophysical modeling, and geological constraints from surface features produces thermal evolutionary models of the martian interior.

### *2.4.1 Isotopic and geologic constraints on thermal models*

Several isotopic results indicate that the martian mantle became heterogeneous within the first 50 Ma of Solar System history and retains that heterogeneity to the present. Isotopic ratios of many short-lived radioactive elements (including $^{182}Hf/^{182}W$, $^{129}I/^{129}Xe$, $^{244}Pu/^{240}U$, and $^{146}Sm/^{142}Nd$) suggest that the martian mantle preserves an early isotopic heterogeneity which is not seen on Earth. Nobel gases, particularly xenon (Xe), indicate an early degassing of the martian atmosphere from the mantle (Hunten *et al.*, 1987; Jakosky and Jones, 1997; Swindle and Jones, 1997; Mathew and Marti, 2001). Strontium (Sr), a lithophile element, shows very little evidence for remixing throughout the history of Mars since its bulk rock values define an approximate 4.5 Ga isochron (Shih *et al.*, 1982; Jagoutz *et al.*, 1994; Borg *et al.*, 1997). The fact that the mantle has not been homogenized during the past ~4.5 Ga indicates that the martian interior has not undergone vigorous convection, unlike Earth where mantle convection drives plate tectonic activity.

Although the isotopic data suggest little to no mantle convection, the long-lived volcanic activity displayed on the surface (Section 5.3.2) argues for mantle convection for most if not all of martian history. While many of the radioisotopes are LILs and concentrate in the crust during differentiation, heating from accretion and core formation are localized within the deep interior. Convection transports heat between regions of differing temperatures. A hot parcel of material is less dense than the surrounding cooler material and, because pressure decreases outward in planetary interiors, will rise, expand, and cool until its temperature and density match the surroundings. Convection removes the heat from the interior and deposits it near the surface where it can escape to space. It is a very efficient mechanism for cooling the interior of a planet.

Volcanic and tectonic activity is primarily centered on the Tharsis Province, an extensive rise covering much of the western hemisphere of Mars (Mouginis-Mark *et al.*, 1992b). Early models suggested that the Tharsis rise resulted from uplift and dynamic support from underlying convection (Banerdt *et al.*, 1992), but MOLA analysis suggests that most of the elevation is the result of accumulation of volcanic materials erupted throughout martian history (Solomon and Head, 1981; Phillips *et al.*, 2001; Smith *et al.*, 2001a), although some mantle support is still required (Kiefer, 2003). Analysis of deep layers exposed in Valles Marineris suggest that volcanism throughout this region peaked early in martian history (McEwen *et al.*, 1999), but crater counts suggest that volcanic activity has occurred until geologically recent times in Tharsis (Hartmann *et al.*, 1999; Neukum *et al.*, 2004). Tharsis also is a major tectonic center, with over half of the features having formed during the earliest period of martian history (Anderson *et al.*, 2001). Models for the long-term maintenance of Tharsis volcanism and tectonism include active convection and mantle plume/superplume activity.

Mantle convection is also suggested by the presence of the martian hemispheric dichotomy. Mariner 9 and Viking Orbiter observations revealed that most of the northern hemisphere is lower and younger than the southern hemisphere (Mutch *et al.*, 1976; Carr, 1981). Explanations for this hemispheric dichotomy included enhanced convection under the northern plains (Wise *et al.*, 1979; Breuer *et al.*, 1993), early plate tectonic activity (Sleep, 1994; Anguita *et al.*, 1998), a single gigantic impact (Wilhelms and Squyres, 1984), or multiple large impacts (Frey and Schultz, 1988). Crustal thickness and topographic information provided by MOLA reveal no correlation between crustal thickness and the topography of the crustal dichotomy nor is the dichotomy boundary circular in form, as would be expected from a single large impact (Zuber *et al.*, 2000). Although buried large impact craters have been revealed by MOLA (Frey *et al.*, 2002) and MARSIS (Watters *et al.*, 2006), their distribution does not reproduce the crustal thickness and topography of the northern plains and dichotomy boundary (Zuber *et al.*, 2000). Impact craters also would be expected to produce an observable gravity signature, which is not observed. Crustal thickness data suggest that the northern hemisphere was thinned by high heat flow early in Mars' history, supporting the theory that differences in mantle convection patterns (perhaps including an early stage of plate tectonic activity) were established early in martian history which gave rise to the hemispheric dichotomy (Zuber *et al.*, 2000; Smith *et al.*, 2001a).

Mars today does not exhibit a magnetic field produced by an active dynamo, but results from the MGS MAG experiment reveal remnant magnetization within ancient crustal rocks (Section 3.5.2), suggesting that Mars had an active dynamo early in its history (Acuña *et al.*, 1998, 1999; Connerney *et al.*, 1999). More recently, weak magnetic signatures have been identified within the buried ancient crust in the northern hemisphere, particularly along the dichotomy boundary (Lillis *et al.*, 2004). The lineated appearance of the magnetized regions suggests possible formation by crustal spreading, similar to the lineated magnetization seen adjacent to ocean rift systems on Earth (Connerney *et al.*, 1999). Cooling of long dike swarms (Nimmo, 2000), accretion of terrains at a convergent boundary (Fairén *et al.*, 2002), and hydrothermal metamorphism (Scott and Fuller, 2004) have also been suggested as explanations for the lineated appearance of these features. Thickness estimates of the magnetized crust range from 18 km, requiring a magnetization of $\sim 25\,\mathrm{A\ m^{-1}}$, to 40 km, with a more reasonable magnetization of $\sim 12\,\mathrm{A\ m^{-1}}$ (Langlais *et al.*, 2004). Regions surrounding the youngest large impact basins (Utopia, Isidis, Hellas, and Argyre) show no evidence of remnant magnetization, suggesting that shock pressures ($>$1–2 GPa) associated with the impact process have demagnetized the surrounding ancient materials (Hood *et al.*, 2003; Rochette *et al.*, 2003). This impact-induced shock demagnetization places constraints on the timing of the magnetic dynamo since the basin ages can be estimated from crater counts (Section 5.1.3). Such analysis suggests that Mars' dynamo operated for no more than the first 500 Ma of the planet's history. Temperature differences between the core and the outer parts of the planet are the likely driving mechanism for the dynamo. Mechanisms proposed for cessation of the dynamo activity include changes in mantle convection (Nimmo and Stevenson, 2000), solidification of the core (Stevenson, 2001), and reduction in core heat production (Williams and Nimmo, 2004).

## 2.4.2 Thermal models for Mars

Thermal evolutionary models of the martian interior are obtained by combining information from isotopic analysis, geologic history, and geophysical modeling. Often the initial data appear contradictory, such as the result from isotopic analysis suggesting little mantle convection over martian history versus the geologic data suggesting long-lived volcanic activity on the planet. The thermal models must be able to explain such contradictions and often evolve over time as new data are acquired and advances in numerical modeling occur.

Accretionary heating, core formation, and decay of short-lived radionuclides provide almost twice as much heat as is necessary to melt the entire planet (Elkins-Tanton *et al.*, 2003, 2005). Thus all current Mars thermal models begin with a magma ocean phase. Core formation appears to require a magma ocean since melt–silicate separation in the FeO-rich martian mantle can only be accomplished when the silicate mantle is molten (Terasaki *et al.*, 2005). One of the early arguments against a martian magma ocean was the lack of a plagioclase-rich crust, as is seen with the anorthositic lunar highlands. However, the mantle of Mars is more water-rich than that of the Moon and water in silicate melt tends to suppress the formation of plagioclase. Also, Mars is larger than the Moon and pressures near the base of the magma ocean could produce majorite and garnet phases, which will sequester the aluminum that would otherwise form a plagioclase-rich crust (Borg and Draper, 2003; Elkins-Tanton *et al.*, 2003; Agee and Draper, 2004).

The depth of the martian magma ocean is not well constrained. Righter *et al.* (1998) used isotopic abundances in melt inclusions from the martian meteorites to estimate a shallow magma ocean of 700–800 km depth while Elkins-Tanton *et al.* (2003) argue that extrapolating terrestrial magma ocean depths (Rubie *et al.*, 2003) would produce a martian magma ocean of at least 1500 km depth. A deep magma ocean requires an insulating atmosphere, presence of a stable solid lid at the surface, or rapid heat production to be maintained against the cooling effects of space (Abe, 1997). Convection, driven by temperature differences within the planet, will homogenize this magma ocean.

Fractional crystallization occurs as the temperature drops and/or pressure increases, causing minerals to crystallize out of a homogeneous melt and produce cumulate layers. Modeling of a 2000-km-thick martian magma ocean (Elkins-Tanton *et al.*, 2003) indicates that majorite and $\gamma$-olivine (ringwoodite) crystallize near the bottom of the magma ocean at pressures between 14 and 24 GPa. Garnet crystallizes between 12.5 and 14 GPa, followed by olivine and pyroxene crystallizing at lower pressures. However, the cumulate stratigraphy produced by fractional crystallization is not stratified according to density – high-density iron-rich and incompatible element-rich minerals form after low-density minerals have crystallized. This unstable density

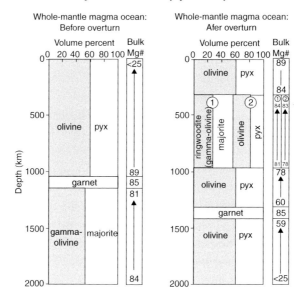

Figure 2.6 The stratigraphy of a mantle-wide magma ocean is shown in this model from Elkins-Tanton *et al.* (2005). The original stratigraphy is not stratified by density, resulting in mantle overturn and a different post-overturn stratigraphy based on density. (Reprinted by permission from American Geophysical Union, Elkins-Tanton *et al.* [2005], Copyright 2005.)

structure causes overturn of the mantle cumulates (Figure 2.6). The rate of overturn is inversely proportional to cumulate layer thickness, with layers >1800km thick overturning within $10^5$ years (Elkins-Tanton *et al.*, 2005). As solidification proceeds, the time required for overturn decreases, thus the fastest overturn occurs near the end of mantle solidification. The time for complete crystallization of Mars' magma ocean has been extrapolated from terrestrial magma ocean calculations and is on the order of a few tens of millions of years (Tonks and Melosh, 1990; Abe, 1997; Solomatov, 2000). The timescales for core formation (10–30Ma: Kleine *et al.*, 2002), mantle differentiation (~30Ma: Harper *et al.*, 1995; Shih *et al.*, 1999), and isotopic heterogeneity of the mantle (~50Ma: Section 2.4.1) are consistent with the timescales for magma ocean crystallization. Cool layers from near the surface will descend to the core–mantle boundary during cumulate overturn, removing heat from the core and shutting off the early magnetic dynamo within 15 to 150Ma after the overturn (Elkins-Tanton *et al.*, 2005).

Whether plate tectonic activity occurred early in martian history is still debated. Early heating and magma ocean formation would favor vigorous mantle convection which could lead to plate tectonic activity once the crust formed. Sleep (1994) proposed that the ancient crust in the northern hemisphere was subducted under the southern highlands and that the current low-lying northern plains were produced by seafloor spreading. The Tharsis volcanoes, in this model, were an island arc volcanic chain produced by subduction. Sleep proposed that plate tectonics operated through the early Hesperian (~3.7 Ga ago, according to Hartmann and Neukum, 2001), but obvious geologic evidence of plate tectonic activity during this period is lacking (Pruis and Tanaka, 1995; Zuber, 2001) and the detection by MOLA and MARSIS of an ancient cratered crust buried beneath the northern plains indicates that any plate tectonic activity must have ceased long before 3.7 Ga ago (Frey *et al.*, 2002; Watters *et al.*, 2006). The martian meteorite isotopic data indicating establishment of mantle heterogeneity by ~50 Ma after Solar System formation also argue against vigorous mantle convection operating until ~3.7 Ga ago.

Current discussion of martian plate tectonics places this activity much earlier, during the period when Mars' magnetic field was active. The lineated pattern of the crustal remnant magnetization in Terra Cimmeria and Terra Sirenum (see Figure 3.8) has been suggested to be direct evidence of plate tectonic activity, forming by crustal spreading similar to that seen on Earth (Connerney *et al.*, 1999, 2005), although other mechanisms also have been suggested (Nimmo, 2000; Scott and Fuller, 2004). Nimmo and Stevenson (2000) find that high surface heat flow, such as would be expected during plate tectonics, helps to drive core convection believed responsible for the magnetic dynamo. Lenardic *et al.* (2004) suggest that formation of the crustal dichotomy directly contributed to the cessation of plate tectonics since formation of the thick southern highlands crust would have insulated the mantle, decreasing the temperature gradient driving mantle convection, and shutting off plate tectonic activity. However, negative buoyancy, necessary for subduction, appears to occur at mantle temperatures much cooler than expected during the early martian history (van Thienen *et al.*, 2004). In addition, Williams and Nimmo (2004) propose that the convection required to drive the magnetic dynamo could have been produced simply by the core being at least 150 K hotter than the mantle immediately after core formation. This hot core scenario is consistent with rapid core formation and mantle isotopic heterogeneity indicated by martian meteorite analysis. It does not require plate tectonics, but predicts that the martian core is still liquid since core solidification would have produced a longer-lived dynamo than is observed. A liquid core is consistent with moment of inertia analysis of the present-day martian interior (Section 3.2.1).

Today Mars is a single-plate (or stagnant lid) planet. If Mars had active plate tectonics during its early history, the thermal evolution must have changed from plate tectonics to stagnant lid at some point. Plate tectonics cools the interior of a planet by transporting that heat to the surface, where it quickly radiates to space through a thin lithosphere. The outer layers of a planet with a stagnant lid are cooled by lithospheric growth, but cooling of the deep interior is inefficient and the lower mantle can remain warm. Crustal thickness, therefore, varies considerably between the two models. The thickness of the current martian crust is estimated to range between 50 km (Zuber *et al.*, 2000) to 200 km (Sohl and Spohn, 1997), with ~100 km being a typical value. A crust no thicker than ~30 km can form even after cessation of plate tectonics according to thermal models of early plate tectonic activity (Breuer and Spohn, 2003). This requires another mechanism to add material to the base of the crust after plate tectonics stops (Norman, 2002) and still produce the geochemical signatures observed in the martian meteorites. Thus, plate tectonics does not seem to be able to explain the observed crustal thickness of present-day Mars.

Breuer and Spohn (2003) have conducted parameterized mantle convection models of stagnant lid versus early plate tectonic scenarios for Mars. Their plate tectonic model predicts a peak in volcanic activity around 2–2.5 Ga ago, which is not supported by the geologic evidence. In addition, their plate tectonic model cannot produce both an early magnetic dynamo and the minimum estimated thickness of the current martian crust. Even adjustments to the temperature and water content of the mantle cannot produce the estimated crustal thickness using the plate tectonic model. Long-lived mantle plumes, such as that proposed for the Tharsis volcanic region, are difficult to produce and maintain in the mantle conditions resulting from early plate tectonics. Breuer and Spohn's models indicate that the present-day crustal thickness, timing of the active magnetic field, geochemical data, declining rate of volcanism, formation of the crustal dichotomy, and prevalence of volcanism in Tharsis are best explained by assuming that Mars has been a one-plate planet throughout its history.

# 3

## Geophysical measurements and inferred interior structure

Geophysical measurements allow scientists to remotely determine the interior structure of a planet (Hubbard, 1984). Gravity deviations from those expected for a homogeneous spherical body provide information about a planet's shape, core, topography, and distribution of subsurface mass. Heat flow measurements provide insights into the thermal history of the planet and the current distributions of radioisotopes. Seismic data reveal the detailed internal structure of the planet and magnetic field data provide information on the core. Thus, geophysical studies provide important constraints on regions of the planet inaccessible to *in situ* study.

### 3.1 Shape and geodetic data

#### 3.1.1 Shape of Mars

The shape of Mars is derived from detailed MOLA topography combined with gravity measurements from MGS Doppler tracking data (Lemoine *et al.*, 2001; Smith *et al.*, 2001a). The shape is defined relative to the planet's center of mass (COM). Mars' equatorial radius is 3396.200km as measured from the COM. COM is offset slightly to the north (primarily because of the Tharsis Bulge), resulting in a north polar radius of 3376.189km, compared to the south polar radius of 3382.580km. Because of its rotation, Mars is slightly flattened (Section 1.4.2), with a value of 0.00648.

The shape of Mars is best approximated as a triaxial ellipsoid (Smith *et al.*, 2001a). We can define a Cartesian coordinate system centered at the COM, with the $z$-axis corresponding to the rotation axis and the $x$- and $y$-axes passing through the equatorial region (Figure 3.1). Usually, but not always, the $x$-, $y$-, and $z$-axes are chosen to correspond to the principal moments of inertia, $A$, $B$, and $C$, respectively. The maximum moment of inertia, $C$, lies along the rotation axis for planets with stable rotation. The moments of inertia are typically selected such that $C > B > A$. The distances between the center and the surface of the reference ellipsoid along each of the principal moments of inertia are designated $a$, $b$, and $c$, respectively, and

Table 3.1 *Mars geodetic parameters from MOLA*

| Triaxial ellipsoid axes lengths | |
|---|---|
| a | 3398.627 km |
| b | 3393.760 km |
| c | 3376.200 km |
| Directions of principal axes | |
| a | 1.0°N  72.45°E |
| b | 0.0°N 324.4°E |
| c | 89.0°N 252.4°E |

*Source*: From Smith *et al.* (2001a).

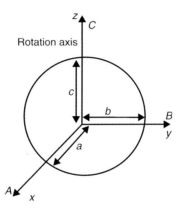

Figure 3.1 Cartesian coordinate system utilized in geophysical analysis. The body's rotation axis corresponds to the *z*-axis in this system. Principal moments of inertia *A*, *B*, and *C* are typically chosen along the *x*-, *y*-, and *z*-axes, respectively. Distances from the center to the surface of the reference ellipsoid along each of the three axes are indicated by *a*, *b*, and *c*.

are called the principal axes of the ellipsoid. The lengths and surface intersection points of Mars' principal axes are given in Table 3.1. The center of the best-fit ellipsoid (center of figure, or COF) is offset from the COM along the polar (*c*) axis by 2.986 km in the direction of the south pole and along the *b*-axis by 1.428 km in the direction of the Tharsis Bulge.

### 3.1.2 Coordinate systems

Locating a feature on the surface requires defining a coordinate (or geodetic) grid. Latitude is measured in two ways because of Mars' oblate shape: aerographic latitude ($\phi$) and aerocentric latitude ($\phi'$). To obtain $\phi$, one draws a best-fit reference sphere centered on the planet's COM (Figure 3.2). A horizontal line tangent to the

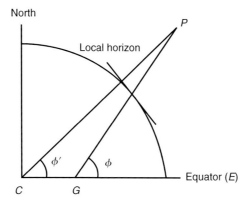

North

Local horizon

P

$\phi'$ $\phi$

C G

Equator (E)

Figure 3.2 Martian coordinate systems are based on both aerographic and aerocentric systems. The reference sphere is the best-fit sphere centered at the planet's center of mass. The aerocentric latitude ($\phi'$) is the latitude measured relative to the center of the reference sphere (C). Because of the oblate shape of the planet, the normal to the local horizon at any point on the surface will not necessarily intersect C, but rather be offset to some point G. The angle PGE is the aerographic latitude ($\phi$).

reference sphere is drawn through the point of interest on the surface. The angle between the equator (E) and the normal to this tangent surface is the aerographic latitude ($\phi$). Because of Mars' oblate shape, this is not the same as the aerocentric latitude ($\phi'$), which is the angle between P, the center of the reference sphere (C), and E. Gravity and topography calculations typically use aerocentric latitude while imaging and mapping (cartography) teams have traditionally used aerographic latitude. However, the coordinates of surface features have been tied to the aerocentric position since MOLA results were first reported. Thus, one must be aware that the coordinates of surface features on Viking-based maps are offset from their positions on recent, MOLA-based maps.

Terra Meridiani, a dark feature easily visible in Earth-based telescopes, was first designated as the martian prime meridian on Beer and Mädler's 1830 map. Subsequent maps refined the location of the 0° longitude line within this region. Following Mariner 9 imaging of Mars, the prime meridian was defined as passing through a 500-m-diameter crater called Airy-0 (de Vaucouleurs *et al.*, 1973).

The 1970 International Astronomical Union (IAU) adopted the convention that longitude, measured from 0° to 360°, should increase in the direction of rotation. For planets rotating directly, like Mars, this would result in longitude being measured eastward from the prime meridian. However, early telescopic observers drew their maps with longitude increasing toward the west so that longitude increased as they observed throughout the night. The west-longitude system, although contrary to that adopted by the IAU, continued to be used through the

Viking missions. Starting with MGS data, however, one has to be careful to note the coordinate system in use, particularly when comparing datasets. MOC imagery utilizes the west-positive coordinate system, but most other datasets use longitude increasing towards the east.

## 3.2 Gravity and topography

If Mars were a homogeneous sphere, the trajectories of orbiting spacecraft could be accurately predicted and would suffer no deviations from the predicted orbit. Not only is Mars differentiated and thus heterogeneous but it also displays variations in topography and mass near the surface. The gravitational effects of these mass anomalies lead to deviations in the expected orbital paths of spacecraft. Analyses of these gravity anomalies provide important insights into the interior structure of Mars.

### 3.2.1 Gravity analysis

The gravitational acceleration ($g$) experienced by a spacecraft orbiting a planet of mass $M$ at a distance of $r$ from the planet's center is

$$\vec{g} = \frac{GM\hat{r}}{r^2} \tag{3.1}$$

where $G$ is the universal gravitational constant. The gravitational acceleration is the integral of the gravitational potential ($U$), therefore:

$$U = -\frac{GM}{r}. \tag{3.2}$$

Since all of the planet's mass lies inside the surface, the gravitational potential exterior to the planet satisfies Laplace's equation:

$$\nabla^2 U_{\text{ext}} = 0. \tag{3.3}$$

The general solution to Laplace's equation in spherical coordinates is (Hubbard, 1984):

$$U(r, \theta, \lambda) = \sum_{l=0}^{\infty} \sum_{m=-l}^{l} [\alpha_{lm} r^l + \beta_{lm} r^{-(l+1)}] Y_{lm}(\theta, \lambda). \tag{3.4}$$

This equation gives the gravitational potential at a particular spot on the planet, where $r$=distance from center, $\theta$ is the colatitude ($=90°$-latitude), and $\lambda$ is the longitude; $\alpha_{lm}$ and $\beta_{lm}$ are constants. In planetary physics applications, $\alpha_{lm}$ is set to zero so that the potential vanishes at infinity. The index, $l$, is the degree of the harmonic and indicates the rate at which the gravitational potential varies in latitude.

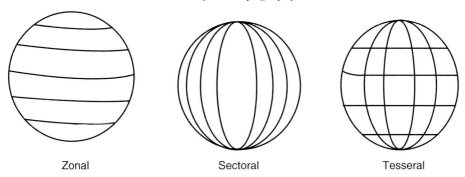

Zonal        Sectoral        Tesseral

Figure 3.3 The expansion of the gravity potential produces three types of harmonics. Zonal harmonics provide information on the mass distribution parallel to the equator while sectoral harmonics provide the longitudinal distributions of mass. Tesseral harmonics subdivide the planet into small blocks to determine the localized distributions of mass.

The index, $m$, is the order and indicates how rapidly $U$ varies in longitude. $Y_{lm}$ are spherical harmonics, defined by

$$Y_{lm} = \sqrt{\frac{(2l+1)}{(4\pi)}} \sqrt{\frac{(l-m)!}{(l+m)!}} P_l^m(\cos\theta)e^{im\lambda}. \tag{3.5}$$

The Legendre polynomials, $P_l^m$, are defined by

$$P_l^m(x) = \frac{(-1)^m (1-x^2)^{m/2}}{2^l l!} \frac{d^{l+m}(x^2-1)^l}{dx^{l+m}}. \tag{3.6}$$

The coefficients $\beta_{lm}$ are associated with the mass distribution inside the planet. Since the potential is measured outside of the planet, the Legendre polynomials are related to the spherical harmonics, and Eq. (3.4) is typically rewritten as

$$U = \left(\frac{GM}{r}\right)\left\{1 + \sum_{l=2}^{n}\sum_{m=0}^{l}\left(\frac{R}{r}\right)^l P_l^m(\cos\theta)[C_{lm}\cos(m\lambda) + S_{lm}\sin(m\lambda)]\right\}. \tag{3.7}$$

The radius of the reference sphere or ellipsoid is $R$ and $n$ is the limiting degree of the expansion. Here $C_{l0}$ are the zonal harmonics, which provide information on mass distributions parallel to the equator (Figure 3.3). The sectoral harmonics, $S_{l0}$, define the mass distributions perpendicular to the equator. Tesseral harmonics ($C_{lm}$ and $S_{lm}$, $m > 0$) further subdivide the planet into smaller blocks, allowing localized variations in mass to be discerned.

Geophysicists often use $J$ instead of the $C_{l0}$ nomenclature for the zonal harmonics:

$$J_l = -C_{l0}. \tag{3.8}$$

If the center of the reference figure coincides with the center of mass of the body, $J_1=C_{11}=S_{11}=0$. Unfortunately for Mars, the COM–COF offset makes these coefficients non-zero (Smith *et al.*, 2001a). $J_2$ is the zonal harmonic representing the largest deviation from a perfect sphere, which results from the planet's equatorial bulge. The principal moments of inertia ($A$, $B$, and $C$) are related to the lower order and degree harmonics by

$$MR^2 J_2 = C - \frac{(A+B)}{2} \tag{3.9}$$

$$MR^2 C_{22} = \frac{(B-A)\cos(2\Delta\lambda)}{4} \tag{3.10}$$

$$MR^2 S_{22} = \frac{(B-A)\sin(2\Delta\lambda)}{4} \tag{3.11}$$

where $\Delta\lambda$ is the difference in longitude between the minimum moment of inertia, $A$, and the $x$-axis.

For planets which rotate "sufficiently rapidly" (typically those which have not been tidally despun), the $B$ and $A$ moments of inertia are much smaller than the maximum moment of inertia, $C$. Thus,

$$|B-A| \ll |C-A| \quad \text{and} \quad |B-A| \ll |C-B|. \tag{3.12}$$

Equation (3.9) can then be rewritten as

$$MR^2 J_2 \cong C - A \cong C - B. \tag{3.13}$$

Thus, by determining $J_2$ from the gravity analysis, one can determine the differences of the principal moments of inertia. To obtain absolute values of the principal moments of inertia, one uses the precession constant ($H$), which can be derived from the rate of precession of the planet's rotation axis:

$$H = \frac{C-A}{C}. \tag{3.14}$$

Combining Eqs. (3.13) and (3.14) allows one to determine $C/MR^2$ and $A/MR^2$. $C/MR^2$ is of particular interest since it indicates the amount of central condensation in a planet. A homogeneous spherical body has $C/MR^2=0.4$. As a body becomes more centrally condensed (i.e., forms a larger core), $C/MR^2$ becomes smaller than 0.4. Mars' $J_2$ varies seasonally due to the sublimation/condensation cycles of $CO_2$ between the polar caps and atmosphere, but the average value from MGS Doppler tracking is $1.96\pm0.69\times10^{-9}$ (Smith *et al.*, 2001b). Comparison of Mars' rotation rate between Viking lander and Mars Pathfinder data gives a precession rate of $-7576\pm 35$ milli-arcseconds per year, leading to a value of $C/MR^2=0.3662\pm0.0017$

(Folkner *et al.*, 1997). This indicates that Mars has a small core, but there is still debate as to whether it is solid (Smith *et al.*, 2001b) or liquid (Yoder *et al.*, 2003).

The primary contributor to $J_2$ is the planet's flattening, producing the equatorial bulge. Since flattening primarily depends on how rapidly the planet is spinning, $J_2$ can be related to the planet's angular rotation rate ($\omega$) through a dimensionless measure of the centrifugal potential ($q$):

$$q = \frac{\omega^2 R^3}{GM} = 2J_2.\tag{3.15}$$

Strictly speaking, Eq. (3.15) is only valid in planets where the interior density is constant. For more realistic planets, where density increases with depth, the coefficient before $J_2$ becomes greater than 2. For bodies in hydrostatic equilibrium, $J_2$ and $q$ can be related to the flattening ($f$) by

$$f \simeq \left(\frac{3}{2}\right)J_2 + \left(\frac{1}{2}\right)q.\tag{3.16}$$

### 3.2.2 Gravity anomalies, isostacy, and crustal thickness

The equipotential surface corresponding to a spherical planet's mean radius ($R$) is called the geoid. Geophysicists define the martian geoid at a radius of 3396 km from COM (Neumann *et al.*, 2004). Variations in gravitational potential from the geoid are called gravity anomalies and represent uneven mass distribution on the surface or within the planet. Gravity anomalies are measured in gals, where 1 gal $= 10^{-2}$ m s$^{-2}$. Most gravity anomalies are very small and are measured in milligals (mgal). Smaller masses can be discerned as the order ($l$) and degree ($m$) increase in Eq. (3.7). Detailed trajectory analyses of the various Mars orbiter missions have led to gravity models up to order and degree 85 (Neumann *et al.*, 2004).

Using Eq. (3.1), we find that the acceleration due to gravity ($g_0$) at the geoid is

$$g_0 = \frac{GM}{R^2}.\tag{3.17}$$

At some height $h$ above the geoid, this becomes

$$g(h) = \frac{GM}{(R+h)^2}.\tag{3.18}$$

Using a Taylor series expansion, Eq. (3.18) becomes

$$g(h) = \frac{GM}{R^2} - \frac{2GM}{R^3}h = g_0 - \frac{2GM}{R^3}h.\tag{3.19}$$

Thus, when a spacecraft is at some height $h$ above the geoid, the gravitational acceleration that it measures is less than that measured at the geoid. The difference in gravity due to elevation above the geoid is called the free air anomaly and the term $-(2GM/R^3)h$ is the free air correction, indicated by $g_{FA}$.

The free air anomaly ignores any mass which exists between the geoid and the spacecraft position at $h$. Thus, if the spacecraft passes over a mountain, it might experience an additional gravitational acceleration due to the mountain's excess mass. In 1749 the French mathematician Pierre Bouguer applied Gauss's law in a gravitational field to an infinite slab of density $\rho$ and thickness $h$ to obtain the Bouguer correction:

$$g_B = 2\pi G\rho h. \qquad (3.20)$$

One final correction occurs for latitudinal variations in $g$ due to rotation and the planet's oblate shape. This latitudinal correction is indicated by $\gamma_0$. The complete description of gravity is called the Bouguer anomaly and is given by

$$g(h) = \frac{GM}{R^2} - \frac{2GM}{R^3}h - 2\pi G\rho h - \gamma_0 = g_0 + g_{FA} - g_B - \gamma_0. \qquad (3.21)$$

Topography and the Bouguer anomaly are often correlated on Mars, providing insights into how topographic features are supported. If the free air anomaly (Eq. 3.19) is approximately zero, the feature is isostatically compensated. Isostacy is the balance between the weight of a crustal block and the buoyant force exerted on it. In Figure 3.4a, we see a crustal block of cross-sectional area $A$ and density $(\rho - \Delta\rho)$ resting in material of density $\rho$. The block straddles the reference surface, extending a height $h$ above the surface and a depth $d$ below it. The upward buoyancy force ($F_B$) on the block is given by Archimedes' principle:

$$F_B = \rho A g d. \qquad (3.22)$$

$F_B$ must be balanced by the block's weight for the block's position to be stable:

$$(\rho - \Delta\rho)(h + d)Ag = \rho A g d. \qquad (3.23)$$

Equation (3.23) reduces to the isostacy equation:

$$\frac{h}{h+d} = \frac{\Delta\rho}{\rho}. \qquad (3.24)$$

A zero free air anomaly (Eq. 3.19) or a negative Bouguer anomaly (Eq. 3.20) indicates complete isostatic compensation. There are two ways in which isostatic compensation can be achieved. Topography which has a constant density but where the depth of the below-surface root is greater than the height of the feature above the reference surface is compensated through Airy isostacy (Figure 3.4b). Alternately, Pratt isostacy (Figure 3.4c) assumes that the depth of compensation is the same for all topographic features but that the density varies, with higher features having lower density. Although Belleguic *et al.* (2005) find density variations across the

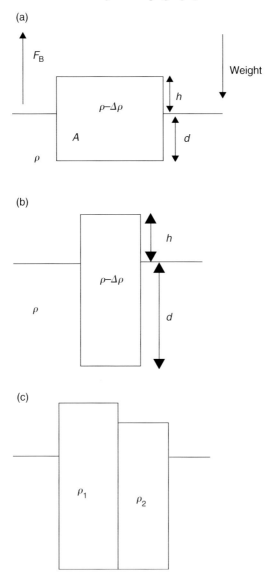

Figure 3.4 Surface mass distributions such as mountains extend below the surface to a depth which depends on the difference in density between the block and underlying material ($\rho - \Delta\rho$ and $\rho$, respectively). (a) The height ($h$) and depth ($d$) of a block of material (with cross-sectional area $A$) relative to the surface is determined by the balance between the block's weight and the buoyant force ($F_B$) from the underlying material. (b) Airy isostacy says that this balance occurs because mountain blocks have roots that extend a greater depth than the mountain's height above the surface. (c) Pratt isostacy has all blocks reaching the same depth, but the above-surface height varies because of variations in block density.

martian surface and suggest that Pratt isostacy might apply, most researchers argue that Airy isostacy is the major compensation mechanism operating on Mars (Zuber *et al.*, 2000; McGovern *et al.*, 2002).

Figure 3.5 shows a Bouguer anomaly map obtained from the MGS MOLA and Radio Science investigations (Neumann *et al.*, 2004). Comparison with the MOLA-derived topography map of Mars (Figure 3.6) shows correlations between gravity and topographic features such as volcanoes and impact basins. Large impact craters and basins typically display a positive Bouguer anomaly because of post-impact uplift of the underlying Moho (the boundary between the crust and the mantle) and infilling by volcanic and sedimentary deposits. Isostatically compensated mountains such as Alba Patera and the Elysium volcanic region display highly negative Bouguer anomalies.

Crustal thickness variations can be modeled using the gravity data, topography, and assumptions about density variations in the crust and mantle (Zuber *et al.*, 2000; Neumann *et al.*, 2004). Crustal thickness varies with latitude, being thicker in the southern hemisphere than in the northern, but also varies with longitude (Figure 3.7). Neumann *et al.* (2004) report an average thickness of 32 km for the northern plains crust versus 58 km for the southern highlands crust. Interesting, the transition in crustal thickness between the two hemispheres does not exactly correlate with the hemispheric dichotomy, arguing against an impact origin for the dichotomy (Zuber *et al.*, 2000).

### 3.2.3 Topography

Resolution of the topographic variations on Mars has been greatly improved by the MOLA experimental results. The laser altimeter provided a vertical accuracy of ~1 m relative to Mars' COM and the topographic grid has a resolution of $1/64°$ in latitude and $1/32°$ in longitude (Smith *et al.*, 2001a). Zero elevation corresponds to the geoid, the equipotential surface corresponding to a distance of 3396 km from COM at the equator. The highest point on Mars is the Olympus Mons volcano summit at 21.287 km and the lowest point is in the Hellas impact basin at −8.180 km. Figure 3.6 clearly shows the hemispheric dichotomy between the low-lying northern plains and the elevated southern highlands.

### 3.3 Seismic data

Mean density and gravity data provide constraints on the interior structure of a planet, but analysis of seismic energy propagation produces the most detailed information on interior heterogeneities. The presence of volcanic and tectonic features on Mars suggests that the planet has been, and may still be, seismically active. Seismic activity likely peaked during the early period of Tharsis Bulge

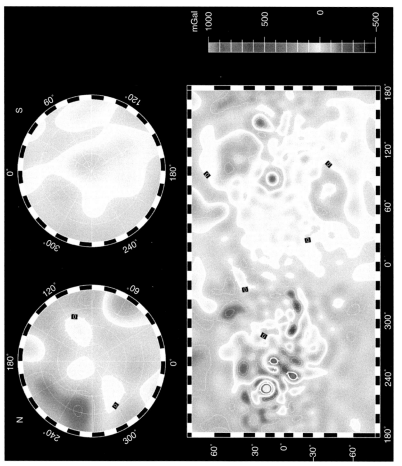

Figure 3.5 Gravity anomaly map, as measured from the vertical accelerations of MGS. A strong positive gravity anomaly is seen between 240°E and 300°E longitude which corresponds to the Tharsis volcanic province, while a strong negative anomaly in the equatorial region between 300°E and 0°E correlates with Valles Marineris. However, many topographic features do not correlate with gravity anomalies. (Image PIA02054, NASA/Goddard Space Flight Center (GSFC).) See also color plate.

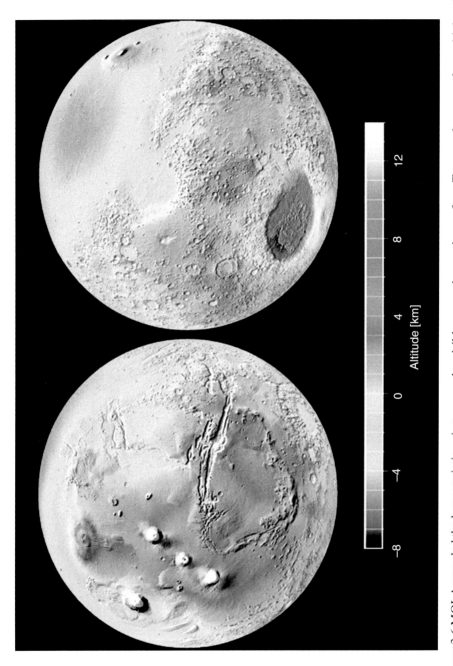

Figure 3.6 MOLA revealed the large variations in topography visible across the martian surface. Topography ranges from a high at the summit of Olympus Mons to a low in the bottom of the Hellas Basin. (Image PIA02820, NASA/JPL/GSFC/MGS MOLA team.) See also color plate.

Figure 3.7 MOLA observations revealed the large variations in crustal thickness across the planet. This image shows the crustal thickness variations (gray band) across a transect from near the north pole across the Elysium volcanic region and into the southern highlands. Thickness of the gray band corresponds to the crustal thickness at each point. (Image PIA00957, NASA/GSFC.)

formation and has declined since then (Golombek *et al.*, 1992). Nevertheless, analysis of cooling rates (Phillips, 1991) and strain rates derived from slip along surface faults (Golombek *et al.*, 1992) suggests that Mars remains seismically active at the present time, with perhaps 14 seismic events of equivalent magnitude 4 or greater occurring each year. Meteorite impact is another source of seismic activity on Mars (Davis, 1993).

Seismic energy is released during planetary response to stress and strain. Energy is distributed both through the interior (body waves) and along the surface (surface waves) of the body. Extremely large seismic events can cause the whole planet to vibrate (free oscillations). Body waves are the most important for determining the internal structure of the planet. The fastest of the body waves, and thus the first to arrive at a seismic station after the event, are compressional waves called primary or P-waves. Secondary or S-waves are distortional, or shear, waves. The velocities of the P- and S-waves depend on the characteristics of the material through which they pass. The bulk modulus ($K$) is a measure of a material's incompressibility and the shear modulus ($\mu$) measures the rigidity of the material to an applied force. If $\rho$ is the density of the material, the P- and S-wave velocities ($V_P$ and $V_S$, respectively) are given by

$$V_P = \sqrt{\frac{K + \left(\frac{4}{3}\right)\mu}{\rho}} \qquad (3.25)$$

$$V_S = \sqrt{\frac{\mu}{\rho}}. \qquad (3.26)$$

$K$ is always greater than zero, so $V_P$ is always greater than $V_S$. Fluids have no rigidity ($\mu=0$), thus S-waves do not travel through liquids.

Mars is differentiated, with denser materials such as iron comprising the core and lower-density materials like silicates constituting the crust (Section 2.2.4). The increase in pressure with depth causes phase transitions among minerals and the increase in temperature towards the core can result in melting of some regions. Thus, the values of $K$, $\mu$, and $\rho$ are constantly changing as one traverses the interior of a planet. Seismic waves also refract as they encounter layers with different material properties. By analyzing the arrival times of P- and S-waves at several surface locations, geophysicists can reconstruct the path taken by and the velocities of the seismic waves to obtain a detailed view of the interior structure of the planet.

Unfortunately, actual seismic data for Mars are non-existent. The Viking landers both carried seismometers, but the Viking 1 instrument failed to deploy and the Viking 2 seismometer was not adequately coupled to the ground, leading to large amounts of noise from lander operations and the wind (Anderson *et al.*, 1977). No marsquakes larger than the magnitude 3 threshold of the seismometers were detected during operation of the Viking 2 seismometer. Penetrators on Mars 96 carried seismometers, but as noted in Section 1.2.2, that mission failed to leave Earth. Seismic network missions have been proposed several times by NASA and ESA, but none of these proposals has been funded to completion (see review in Lognonné, 2005). Until such a seismic network is established, our understanding of the interior structure of Mars will be limited to less detailed information obtained largely through gravity analysis.

## 3.4 Heat flow

Surface heat flow measurements provide constraints on the distribution of heat-producing radioisotopes and the rate of mantle convection. Heat flux ($Q$) is the amount of heat flowing through a particular area in a certain amount of time. The two major mechanisms of heat flow in solid planets like Mars are conduction and convection. Conduction transfers energy through crystal lattice vibrations while convection involves the physical movement of materials at different temperatures. Conduction is the primary mechanism of heat transport within the lithosphere, the outer rigid layer of a planet composed of the crust and upper mantle. Convection occurs in the warmer and more deformable lower mantle, also called the astheno-sphere. Convection likely also occurs within the core, particularly if the core is still molten. Convection occurs not only in fluids but also solids, which can deform under high pressure and temperature conditions. Convection in solids is called solid-state convection.

### 3.4.1 Conduction

The change in temperature ($T$) as a function of depth ($z$) measured towards the center of the planet is called the thermal gradient ($\partial T/\partial z$). In regions where conduction dominates, heat flow is related to the thermal gradient by Fourier's law:

$$Q = k(\partial T/\partial z). \tag{3.27}$$

The thermal conductivity ($k$) is related to the density ($\rho$), specific heat at constant pressure ($c_p$), and thermal diffusivity ($\kappa$) of the material:

$$k = \rho c_p \kappa. \tag{3.28}$$

The rate at which layers gain (or lose) heat is given by

$$\rho c_p \frac{\partial T}{\partial t} = \frac{\partial Q}{\partial z}. \tag{3.29}$$

Combining this with Eq. (3.27) gives the thermal diffusion equation, also called the heat conduction equation:

$$\frac{\partial T}{\partial t} = \kappa \frac{\partial^2 T}{\partial z^2}. \tag{3.30}$$

### 3.4.2 Convection

Convection is much more efficient than conduction at removing internal heat from a planet and dominates in the deeper parts of a planet's interior. Convection occurs when the thermal gradient exceeds some critical value, causing hot material to rise and cooler material to sink. This occurs when material is hotter and therefore less dense than its surroundings, causing the hotter material to be buoyant. As the mass of hotter material rises, it expands and begins to cool. Under adiabatic conditions, it will continue to rise, expand, and cool until its temperature matches that of the surroundings. Convective motions are governed by the conservation equations combined with fluid dynamical equations (e.g., the Navier–Stokes equation). The full description of convection involves finite element modeling and is beyond the scope of this discussion. Interested readers are referred to Turcotte and Schubert (2002).

The dimensionless Rayleigh number ($R_a$) is used to determine if convection will occur. Assume we have a region of the mantle of thickness $d$ which is warmer than its surroundings by a factor of $\Delta T$. If the mantle has a density $\rho_m$, thermal expansion coefficient $\alpha$, thermal diffusivity $\kappa$, and viscosity $\eta$, $R_a$ can be computed from

$$R_a = \frac{\rho_m g \alpha \Delta T d^3}{\kappa \eta} \tag{3.31}$$

where $g$ is the acceleration due to gravity. Convection will occur if $R_a$ is greater than some critical value, which is approximately 1000.

### 3.4.3 Martian heat flux

None of the landed spacecraft has measured surface heat flow on Mars, so all heat flux estimates are based on modeling. While conduction is expected to be the dominant mechanism of heat transport through the lithosphere, convection should occur in the deep martian interior. Mantle convection can be driven by bottom heating, where heat from the core drives the convection, or from internal heating caused by the decay of radioactive elements within the mantle itself. Kiefer (2003) argues that the broad topographic rises associated with Tharsis and Elysium result from internal heating, which produces broad convective upwellings. Mantle plumes, responsible for the individual volcanic constructs, can be embedded within these broad upwelling zones.

McGovern *et al.* (2002, 2004a) used MGS gravity and topography data to estimate the thickness of the elastic lithosphere at the time of surface loading and from that determine the surface heat flux. They estimate that the oldest surface units were emplaced when heat flux was >40 milliwatts per square meter (mW m$^{-2}$) and that $Q$ has declined to values typically <20 mW m$^{-2}$ in recent times (McGovern *et al.*, 2004a). These values are consistent with an estimated $Q$ of ~37 mW m$^{-2}$ near 3.0–3.7 Ga ago based on the spacing of tectonic wrinkle ridges (Montési and Zuber, 2003). However, Grott *et al.* (2005) suggest $Q$ between 54 and 66 mW m$^{-2}$ between 3.5 and 3.9 Ga ago based on elastic thickness estimates in the Coracis Fossae region. Heat flux likely varies with location on Mars, as it does on Earth, so the variations in time and geographic position suggested by these different studies may not be inconsistent. Improved constraints on heat flux will have to await emplacement of a network of seismic and geothermal measurement stations across the martian surface.

## 3.5 Magnetics

Detection of an active magnetic field or remnant magnetization in crustal rocks indicates that the planet contains an interior conductive layer. The conductive layer in terrestrial planets is a metallic iron-rich core, probably mixed with some sulfur. The core must be at least partially liquid for production of an active magnetic field. As long as the core is composed of material which conducts electricity, it can produce a time-varying electric field which induces a magnetic field. The area of physics which studies how motions within a fluid core set up the electric currents which produce planetary magnetic fields is called magnetohydrodynamics (MHD) or magnetic dynamo theory (Stevenson, 2003).

### 3.5.1 Active dynamo

The basis of MHD analysis is Maxwell's equations. Faraday's law of inductance relates the electric ($E$) field to a time-varying magnetic ($B$) field:

$$\vec{\nabla} \times \vec{E} = -\frac{\partial B}{\partial t}. \tag{3.32}$$

Ampère's law describes how currents ($J$) and time-varying electric fields produce a magnetic field:

$$\vec{\nabla} \times \vec{B} = \mu_0 J + \mu_0 \varepsilon_0 \frac{\partial E}{\partial t} \tag{3.33}$$

where $\mu_0$ is the magnetic permeability of free space ($=4\pi \times 10^{-7}$ N A$^{-2}$) and $\varepsilon_0$ is the permittivity of free space ($=8.85 \times 10^{-12}$ C$^2$ N$^{-1}$ m$^{-2}$). In most planetary magnetic fields, $\mu_0 \varepsilon_0 (\partial E / \partial t)$ is very small and can be ignored. The current ($J$) induced by the magnetic field is given by Ohm's law:

$$\vec{J} = \sigma[\vec{E} + (\vec{v} \times \vec{B})] \tag{3.34}$$

where $\sigma$ is the electrical conductivity of the material and $v$ is the velocity of the conducting fluid. Combining Eqs. (3.32), (3.33), and (3.34) gives the induction equation:

$$\frac{\partial \vec{B}}{\partial t} = \vec{\nabla} \times (\vec{v} \times \vec{B}) + \left(\frac{1}{\mu_0 \sigma}\right) \nabla^2 \vec{B}. \tag{3.35}$$

Equation (3.35) shows how the magnetic field $B$ varies with time. The first term on the right shows that the conducting fluid must be moving with velocity $v$ in order to maintain the magnetic field. The second term on the right is called the diffusion term and $1/(\mu_0 \sigma)$ is the magnetic diffusivity. In the absence of fluid motions (i.e., $v=0$), the magnetic field decays over some time period related to the magnetic diffusivity. For terrestrial planets, this decay period is of order $10^4$ years. Since this is much less than the ~4.5 Ga age of the Solar System, planets with active magnetic fields such as Earth must be generating their fields at the present time.

During aerobraking, MGS's Magnetometer/Electron Reflectometer (MAG/ER) experiment became the first instrument to obtain magnetic field observations below the planet's ionosphere. The maximum magnitude of Mars' dipole is ~$2 \times 10^{17}$ A m$^{-1}$, corresponding to a magnetic field strength of 0.5 nT at the equator (Acuña *et al.*, 2001). This small value indicates that Mars does not currently have an active magnetic dynamo.

### *3.5.2 Remnant magnetization*

Rocks containing magnetic minerals can retain a remnant magnetization indicative of the strength and polarity of the magnetic field that was present when the rock solidified. Like a compass needle, magnetic minerals align with the direction of the magnetic field lines. As long as the material remains molten, the minerals are free to realign with any changes in the magnetic field direction. The motion of the magnetic minerals is reduced as the temperature drops and the magma begins to solidify. Once the temperature has dropped below a specific value, called the Curie temperature ($T_C$), the magnetic minerals can no longer move and the direction of the magnetic field at that time is "frozen into" the rock, giving rise to thermal remnant magnetization (TRM). Minerals that retain the ability to exhibit TRM over geologically long time periods are called ferromagnetic minerals. Ferromagnetism results from the spin and orbital angular momenta of the electron, which give rise to a magnetic dipole for the electron. In atoms with filled electron shells, the spins of the electrons are in up/down pairs resulting in a zero net dipole moment. However, ferromagnetic minerals are composed of atoms with unfilled electron shells. In this case the electrons can align along the direction of the external (planetary) magnetic field and retain that direction after the temperature drops below $T_C$. The Curie temperature varies with mineral composition: $T_C$ of metallic iron is ~1040 K while that of magnetite ($Fe_3O_4$) is ~850 K.

Crustal remnance also can be produced at low temperatures through chemical remnant magnetization (CRM) of paramagnetic minerals. Paramagnetic minerals display magnetic properties when subjected to an external magnetic field, but do not retain this magnetization when the field is removed. Paramagnetic minerals must grow to large grain sizes to obtain CRM, but have difficulty retaining a strong magnetization after the magnetizing field is removed (Connerney *et al.*, 2004).

Titanomagnetite and pyrrhotite in the martian meteorites retain small amounts of remnance, but the long-term stability of this remnance is questioned (McSween, 2002). The best evidence of an ancient magnetic dynamo comes from the MGS MAG/ER discovery of remnant magnetization in some regions of the martian crust (Acuña *et al.*, 1999; Connerney *et al.*, 1999) (Figure 3.8). The strongest regions of remnant magnetization are in the Terra Cimmeria and Terra Sirenum region of the ancient southern highlands (30°–90°S 130°–240°E), with inferred crustal magnetizations of ~10–30 A m$^{-1}$ (Connerney *et al.*, 1999; Langlais *et al.*, 2004). This is an order of magnitude higher than the strongest terrestrial magnetizations. Weaker magnetizations exist throughout most of the southern highlands and in some places under the northern plains (Acuña *et al.*, 1999; Connerney *et al.*, 2005). Regions surrounding large impact basins are weakly magnetized or show no magnetic remnance, probably resulting from demagnetization of rocks containing pyrrhotite

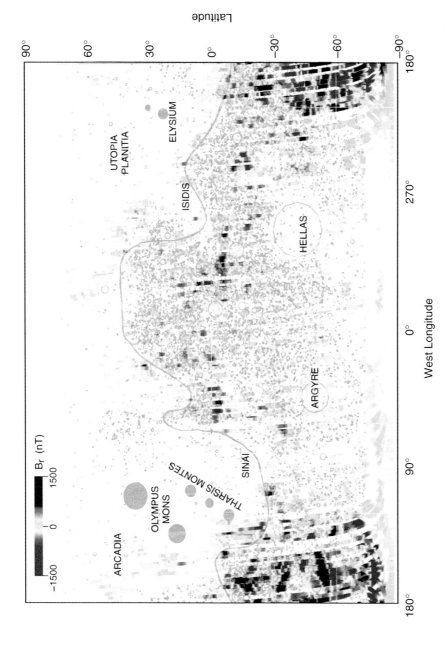

Figure 3.8 MGS's MAG/ER experiment revealed remnant magnetization within ancient rocks across the martian surface. The areas of the strongest magnetizations (red/blue colors) correspond to ancient rocks in the Terra Cimmeria and Terra Sirenum regions. Open circles indicate impact craters and filled circles are volcanoes. The dichotomy boundary is indicated by the solid line. (Image PIA02059, NASA/GSFC/MGS MAG/ER team.) See also color plate.

($Fe_7S_8$), magnetite ($Fe_3O_4$), hematite ($Fe_2O_3$), and/or titanohematite ($Fe_{2-x}Ti_xO_3$) which are exposed to shock pressures $>1 GPa$ (Rochette *et al.*, 2003; Kletetschka *et al.*, 2004). If these basins had formed while the magnetic dynamo was active, these areas would have been remagnetized. The lack of such remagnetization suggests that the dynamo ceased operation before the formation of these large basins $\sim 4$ Ga ago (Acuña *et al.*, 1999). Alternately Schubert *et al.* (2000) suggest that the basins formed prior to the onset of the dynamo, with the magnetization resulting from localized heating and cooling events after basin formation. However, most of the evidence from both geochemical and geophysical arguments supports the idea of an early dynamo (Connerney *et al.*, 2004).

Figure 3.8 shows that crustal remnance appears in lineated patterns with alternating polarities. Connerney *et al.* (1999) noted the similarities in magnetic pattern to terrestrial seafloor spreading and proposed that the martian lineated magnetic anomalies recorded ancient plate tectonic activity in a reversing dipolar field. Faults running parallel to the directions of the magnetic anomalies have been interpreted as transform faults that were active early in martian history (Connerney *et al.*, 2005). Alternately, the linear magnetic anomalies have been explained by magnetite-rich dike intrusions (Nimmo, 2000), accretion of terrains along a convergent boundary (Fairén *et al.*, 2002), and thermal decomposition of iron-rich carbonates into magnetite (Scott and Fuller, 2004). Reconstruction of paleomagnetic pole positions from the magnetic anomalies suggests some clustering and evidence of a reversing dipolar field (Arkani-Hamed, 2001). However, the positions of these poles vary depending on the magnetic anomalies selected. Sprenke and Baker (2000) primarily utilized the Terra Cimmeria and Sirenum anomalies to place the south magnetic pole near 15°S 45°E. Arkani-Hamed (2001) modeled ten isolated magnetic anomalies and found a south magnetic pole position near 25°N 230°E. Hood and Zakharian (2001) used magnetic anomalies in the northern hemisphere to locate the south magnetic pole near 38°N 219°E. All of these pole locations are $>50°$ from the current rotational pole. Most magnetic axes are offset $<15°$ from the planet's rotation axis (Uranus and Neptune are exceptions), leading Sprenke and Baker (2000) and Hood and Zakharian (2001) to propose that either plate motions or reorientation of the planet due to the Tharsis uplift ("polar wander") has moved the paleomagnetic pole positions far from the current geographic poles.

## 3.6 Interior structure of Mars

The mean density of Mars combined with geochemical (mainly martian meteorite analysis) and geophysical (particularly from gravity, topography, and magnetic data) analysis have provided new constraints on the interior structure of Mars. Mars is differentiated into a crust, mantle, and core (Figure 3.9). Gravity and topography

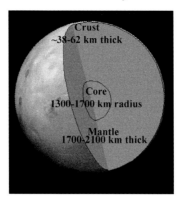

Figure 3.9 Mars' interior structure is inferred from geophysical measurements. The crust varies in thickness between ~38 and 62 km. The mantle is 1700–2100 km thick and the core has a radius of 1300–1700 km. (Background image PIA00974, NASA/MPF team.)

analysis indicates that the crust, composed primarily of basaltic materials (Zuber, 2001) (Section 4.3.2) with a mean density of 2900 kg m$^{-3}$ (Zuber *et al.*, 2000; McGovern *et al.*, 2002, 2004a), is thicker in the southern hemisphere than in the north but also displays substantial regional variations (Zuber *et al.*, 2000; Neumann *et al.*, 2004; Wieczorek and Zuber, 2004) (Figure 3.7). Various techniques have been utilized to estimate crustal thickness on Mars, but all are consistent within the range of 38 to 62 km for the mean thickness (Wieczorek and Zuber, 2004).

The mantle, composed primarily of olivine ($(Mg, Fe)_2SiO_4$) and spinel (the high-pressure polymorph of olivine) extends to a depth of ~1700–2100 km (Zuber, 2001). An FeS-rich core with a radius of ~1300–1700 km lies at the center of the planet. The physical state of the core is uncertain: both solid (Smith *et al.*, 2001a) and liquid (Yoder *et al.*, 2003) have been proposed. The lack of a presently active magnetic dynamo indicates either that the core is not undergoing vigorous convection (Stevenson, 2001) or that solidification of a liquid core is not occurring (Schubert *et al.*, 2000).

The crust and upper mantle constitute the rigid outer layer of the planet called the lithosphere. Deformation in the lithosphere is dominated by brittle failure, creating the tectonic features observed on the martian surface. The lower part of the mantle comprises the asthenosphere, where ductile deformation dominates. The depth where the brittle-to-ductile deformation transition occurs defines the lithosphere–asthenosphere boundary. McGovern *et al.* (2004a) find that the lithospheric thickness varies from <12 km up to as much as 200 km. Conduction is the major mechanism of heat transfer within the lithosphere while convection is expected to dominate in the asthenosphere.

# 4

# Surface characteristics

The surface of Mars displays considerable variability in albedo, composition, surface roughness, and physical properties. These surface characteristics are dictated largely by Mars' bulk composition (Section 2.3) and the geologic processes which have operated over the planet's history (Section 5.3). Spectroscopic observations, radar reflection, and surface operations have provided constraints on martian surface characteristics.

## 4.1 Albedo and color

### 4.1.1 Albedo

Albedo measures the fraction of sunlight reflected by a body. A perfectly reflecting body has an albedo of 1.0 while a perfectly absorbing body has zero albedo. Two common types of albedo which are used in planetary studies are the bond albedo ($A_b$) and the geometric albedo ($A_0$). The value of $A_b$ depends on the incident solar energy and how much of that flux the planet reflects. The solar flux ($F$) at some distance $r$ from the Sun is related to the Sun's luminosity ($L_{solar}=3.9\times10^{26}$ W):

$$F = \frac{L_{solar}}{4\pi r^2}.$$ (4.1)

Because only one side of the planet (radius$=R$) receives this sunlight, the incident flux ($F_i$) on the planet's surface is

$$F_i = (\pi R^2)\frac{L_{solar}}{4\pi r^2} = \frac{L_{solar}R^2}{4r^2}.$$ (4.2)

$A_b$ is the ratio of the reflected flux ($F_r$) to the incident flux ($F_i$), therefore

$$F_r = A_b F_i = \frac{A_b L_{solar}R^2}{4r^2}.$$ (4.3)

However, the amount of sunlight reflected by the body and received on Earth varies with the phase angle $\Phi$ (the Sun–object–Earth angle). $A_0$ is defined as the albedo at $\Phi=0°$. $A_b$ and $A_0$ are related to each other through the phase integral ($q_{ph}$)

$$A_b \equiv A_0 q_{ph}. \tag{4.4}$$

The phase integral gives information on how $F_r$ varies with $\Phi$:

$$q_{ph} \equiv 2 \int_0^\pi \frac{F_r(\Phi)}{F_r(\Phi = 0°)} \sin \Phi d\Phi. \tag{4.5}$$

Average $A_b=0.250$ and $A_0 \approx 0.150$ for Mars, but the regional reflectivity of the martian surface varies with location (Figure 4.1). The darkest regions have $A_0 \approx 0.10$ while $A_0 \approx 0.36$ in the brightest areas (exclusive of the polar caps). The albedo contrast is greatest when Mars is observed through a red filter (centered near 0.6 µm wavelength) and almost disappears when using a blue filter (centered near 0.4 µm). This contrast varies seasonally as dust is redistributed by changing wind directions associated with the alternating recession and expansion of the polar caps. The names given to albedo features by Schiaparelli on his 1877 map are generally still used today for the corresponding topographic features. For example, the high-albedo feature called Hellas is a large impact basin called Hellas Planitia, and the low-albedo feature Syrtis Major is now associated with the volcanic plateau of Syrtis Major Planum.

### 4.1.2 Color

Mars displays a reddish-orange color to the naked eye, but detailed observations show a range in color. The high-albedo surface regions (exclusive of the polar caps) appear reddish-ochre in color while the lower-albedo regions appear grayish. Color views of the planet (Figure 4.2) reveal three discrete units: bright red-ochre, dark gray, and intermediate. The intermediate units appear to be a mixture of the material comprising the bright and dark regions. Color contrast can be enhanced by ratioing the reflectivities obtained using different color filters (James *et al.*, 1996; Bell *et al.*, 1997). These color ratios indicate that the darker units can be subdivided into redder (red/violet ratio $\approx$ 3) and less red (red/violet ratio $\approx$ 2) units. Redder dark units include the Tharsis volcanic region and the southern hemisphere highland plateau regions, while less red dark units are correlated with the intermediate aged ridged plains (Soderblom, 1992). These color variations result from differences in both composition and grain size. The brighter regions are generally correlated with dust deposits while the darker regions correspond to basaltic-rich materials.

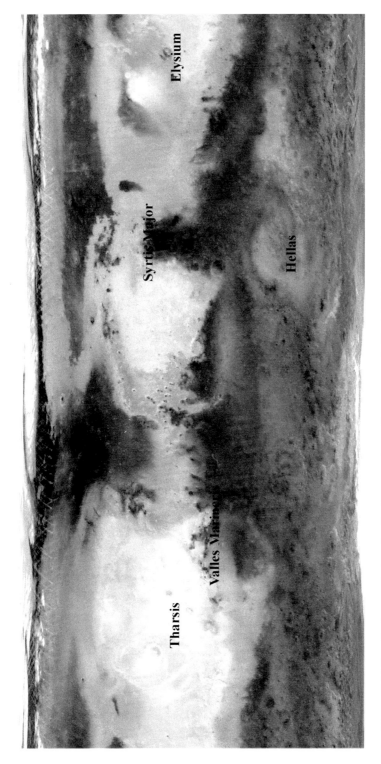

Figure 4.1 Mars displays variations in albedo across its surface, as revealed by MGS TES (Christensen *et al.*, 2001a). (NASA/Arizona State University (ASU).)

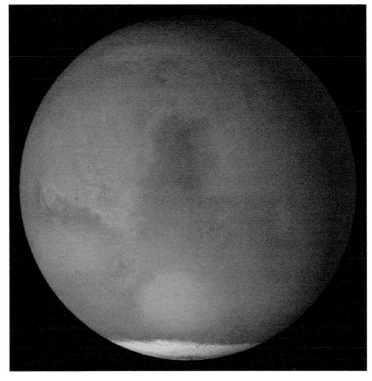

Figure 4.2 Mars displays slight variations in color, as demonstrated in this MOC image taken at $L_S=211°$ (northern autumn/southern spring) in May 2005. The bright south polar cap is visible at the bottom of the image. The dark feature in the center is the volcanic province Syrtis Major Planum and the brighter circular feature below Syrtis Major is the Hellas impact basin. (MOC Release MOC2-1094; NASA/MSSS.) See also color plate.

## 4.2 Surface roughness and structure

The first radar echoes from Mars were obtained in 1963 and the technique has been used during most subsequent oppositions (Simpson *et al.*, 1992). The round-trip echo flight time provides information on the distance between Earth and Mars and topographic variations on the martian surface. The dispersion of the returned echo provides an estimate of surface roughness and the strength of the echo constrains the reflectivity of the surface. The polarization of the returned signal allows estimation of small-scale surface structure. Thus, elevations, slopes, textures, and material properties can be obtained through radar observations.

Ground-based radar observations revealed the diversity of terrains on Mars, including the icy nature of the polar caps, large regions covered by dust mantles, rough lava flows, and indications that lava flows have infilled at least some channels (Simpson *et al.*, 1992; Harmon *et al.*, 1999). A region largely corresponding to the

Medusae Fossae Formation southwest of Tharsis displays almost no radar return (Muhleman *et al.*, 1995; Harmon *et al.*, 1999). Radar data of this "Stealth" region provide evidence that this region is covered by thick layers of fine-grained material (Edgett *et al.*, 1997). Radar data have been used to constrain surface material properties at proposed landing sites and complement the information obtained from orbiting spacecraft on surface roughness, rock abundances, and dust coverings (Section 4.4.2) (e.g., Golombek *et al.*, 2003). These predictions are generally consistent with the actual conditions encountered at the landing sites (e.g., Golombek *et al.*, 1999b, 2005).

Decimeter-scale roughness variations are obtained from ground-based radar (Harmon *et al.*, 1999), and are complemented by kilometer-scale surface roughness estimates from MOLA (Figure 4.3). MOLA reveals that the southern highlands are rougher than the northern plains, largely resulting from the southern hemisphere's heavily cratered nature. The smoothness of the northern plains reflects the sedimentary mantle which overlies this region (Kreslavsky and Head, 2000).

Details about the subsurface properties and structure to a depth of 3–5 km are being revealed by the MARSIS ground-penetrating radar (Picardi *et al.*, 2005; Watters *et al.*, 2006). Attenuation of the radar return provides constraints on the dielectric constant of the subsurface materials, which allows mapping of compositional variations. Preliminary results from the north polar cap are consistent with $H_2O$-ice comprising the summer (remnant) cap and much of the polar layered deposits. MARSIS has revealed the thickness of these polar deposits with a sharp contrast in radar reflectivity corresponding to the base of the layer near 1.8 km (Figure 4.4a). Smaller variations in subsurface properties and compositions are being revealed by MRO's higher-resolution SHARAD ground-penetrating radar (Figure 4.4b) (Seu *et al.*, 2004).

## 4.3 Crustal composition

The compositional diversity of the martian crust has been revealed through remote sensing observations, analysis of martian meteorites, and *in situ* investigations. Most of the compositional information has been obtained through various types of spectroscopic analyses.

### 4.3.1 Methods of compositional analysis

Remote sensing data include observations from Earth-based telescopes, Hubble Space Telescope (HST), and Mars orbiting spacecraft. All of these techniques utilize reflection spectroscopy to determine compositional information. Sunlight consists of a blackbody spectrum modified with absorption lines created by gases in the Sun's

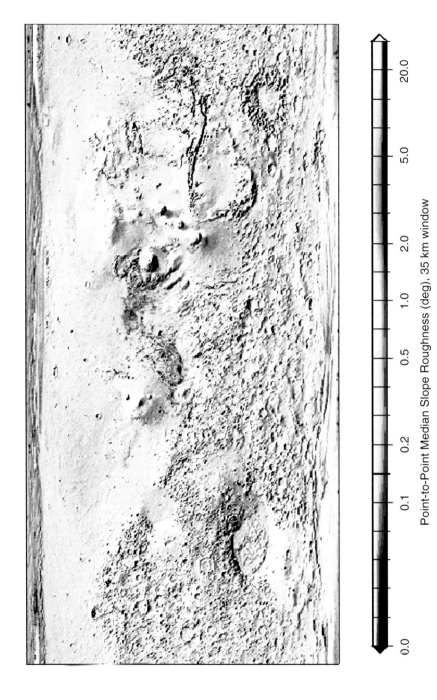

Point-to-Point Median Slope Roughness (deg), 35 km window

Figure 4.3 MOLA analysis provides estimates of the roughness of the martian surface. The southern highlands are rougher than the northern plains. (Image PIA02808, NASA/GSFC.) See also color plate.

(a)

(b)

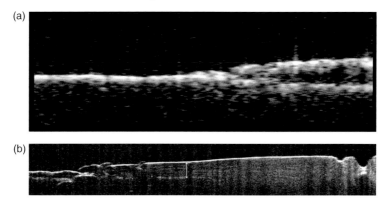

Figure 4.4 The ground-penetrating radars on MEx and MRO provide information
on compositional variations within the near-surface. (a) This MEx MARSIS profile
extends from the northern plains (left) to the north polar cap (right). The split in the
radar return on the right demonstrates the thickness of the polar cap. (ASI/ESA/
University of Rome.) (b) MRO's Shallow Radar (SHARAD) reveals fine layering
within the south polar layered deposits. This profile is 650 km long and the arrow
length corresponds to an 800 m thick segment. The bottom of the arrow indicates
the base of the polar layered deposits. (NASA/JPL-Caltech/ASI/University of
Rome/Washington University in St. Louis.)

photosphere. As sunlight is reflected off of a planetary surface, minerals comprising
its surface will absorb additional wavelengths of energy due to vibrational motions
within the crystal lattices (Christensen *et al.*, 2001a). The wavelengths of these
absorption lines are diagnostic of the minerals doing the absorbing. Removing the
solar spectrum and atmospheric absorptions from the reflected light provides us with
the absorptions due to the planet, which, when compared with laboratory spectra,
allow us to constrain the mineral composition of the surface (Figure 4.5). Many of the
important mineral and atmospheric absorptions occur in the infrared (IR) (Hanel *et al.*,
2003), where the energy has frequencies between $3 \times 10^{11}$ Hz to $3 \times 10^{14}$ Hz, or
wavelengths between 1 μm and 1 mm. Infrared is subdivided into near-IR ($\sim$0.7 to
5 μm), mid-IR (5 to $\sim$30 μm), and far-IR ($\sim$30 to 350 μm). Infrared energy between 5
and 100 μm results from the body's heat and is therefore called thermal IR. Infrared
observations are often reported in wavenumbers rather than wavelengths or frequencies. Wavenumber is the inverse of wavelength and its units in planetary studies
are usually $cm^{-1}$. Thus IR energy with a wavelength of 10 μm corresponds to a
wavenumber of $10^3$ $cm^{-1}$.

Earth-based IR observations are limited to the near-IR because our atmosphere
absorbs the longer wavelengths. The Near Infrared Camera and Multi-Object
Spectrometer (NICMOS) instrument on Hubble Space Telescope (HST) also is
limited to the near-IR, observing in the 0.8 to 2.5 μm region. Mars-orbiting
spacecraft obtain observations over a wider range of wavelengths and produce

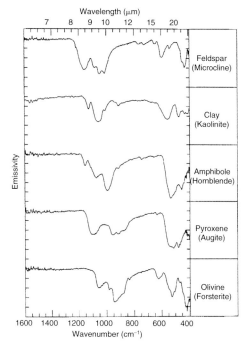

Figure 4.5 Laboratory spectra of several geologically important minerals are shown in this image. Differences in band shape and the wavelengths of specific absorptions distinguish one mineral from another and allow mineralogical analysis from remote sensing platforms. (Reprinted by permission of the American Geophysical Union, Christensen *et al.* [2000b], Copyright 2000.)

higher-resolution data due to their proximity to Mars. Table 4.1 lists the wavelength ranges covered by spectrometers on several of the Mars missions. Many of these spectrometers observe only specific bands within the wavelength range, selected to enhance identification of minerals which are of particular interest. For example, THEMIS investigates the martian surface using IR bands centered at 6.78, 7.93, 8.56, 9.35, 10.21, 11.04, 11.79, 12.57, and 14.88 μm (Christensen *et al.*, 2004a). These bands were selected because absorptions from various silicates (including feldspars and pyroxenes), salts, and carbonates are located near these wavelengths.

Reflection spectroscopy is the major technique utilized in remote sensing observations (Clark and Roush, 1984), but it is not the only way to obtain compositional information. MO's GRS system consists of a gamma-ray spectrometer (GRS), a neutron spectrometer (NS), and a high-energy neutron detector (HEND) (Boynton *et al.*, 2004). GRS detects gamma rays emitted through decay of radioactive elements such as potassium (K), uranium (U), and thorium (Th) as well as gamma rays induced by cosmic rays interacting with non-radioactive elements such as chlorine

Table 4.1 *Wavelengths of observations by selected spacecraft missions*

| Mission/instrument | Wavelengths |
|---|---|
| *Mariner 6/7* | |
|   UV Spectrometer | 110–430 nm |
|   IR Spectrometer | 1.9–14.3 μm |
| *Mariner 9* | |
|   UV Spectrometer | 110–352 nm |
|   IRIS (Infrared Interferometer Spectrometer) | 6–50 μm |
| *Viking 1/2 Orbiter* | |
|   IRTM | 15 μm |
| *Phobos 2* | |
|   ISM | 0.7–3.2 μm |
|   KRFM | 0.3–0.6 μm |
|   Thermoscan | |
|     Visible | 0.5–0.95 μm |
|     Infrared | 8.5–12 μm |
| *Mars Global Surveyor* | |
|   TES | |
|     Spectrometer | 6.25–50 μm |
|     Bolometer | 4.5–100 μm |
|     Albedo | 0.3–2.7 μm |
| *Mars Odyssey* | |
|   THEMIS | |
|     Infrared | 6.62, 7.78, 8.56, 9.30, 10.11, 11.03, 11.78, 12.58, 14.96 μm |
|     Visible | 423, 553, 652, 751, 870 nm |
| *Mars Express* | |
|   OMEGA | |
|     Near-IR | 1.0–5.2 μm |
|     Visible | 0.5–1.1 μm |
|   PFS | 1.2–5 μm |
|   SPICAM | |
|     Ultraviolet | 118–320 nm |
|     Infrared | 1.0–1.7 μm |
| *Mars Reconnaissance Orbiter* | |
|   CRISM (Compact Reconnaissance Imaging Spectrometer for Mars) | 0.362–3.92 μm |

(Cl), iron (Fe), and carbon (C). NS and HEND detect thermal ($<0.4$ eV), epithermal (0.4 eV to 0.7 MeV), and fast (0.7–1.6 MeV) neutrons emitted through cosmic ray interactions with surface minerals. These neutrons can pass through the surface largely unaffected or can be absorbed, depending on surface composition. Comparison of thermal, epithermal, and fast neutron fluxes emitted from the surface provides constraints on the composition within ~1m of the surface. For example, the

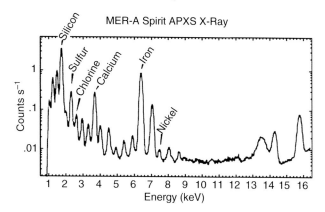

Figure 4.6 Data from Spirit's APXS instrument reveal emissions associated with specific elements. The elements identified in this spectrum have been determined primarily from the X-ray analysis. (Image PIA05114, NASA/JPL/Max-Planck Institute for Chemistry.)

presence of hydrogen (H) in the surface, typically found in the form of $H_2O$ in terrestrial planets, will absorb epithermal neutrons but not thermal neutrons. Carbon dioxide ($CO_2$) ice will allow both thermal and epithermal neutrons to pass. Thus the low flux of epithermal neutrons and high flux of thermal neutrons in the martian polar regions during the summer suggest the presence of $H_2O$ rather than $CO_2$ ice.

The rovers on MPF and MER each carried an Alpha Particle X-Ray Spectrometer (APXS) to determine elemental compositions of surface rocks and soil (Rieder *et al.*, 1997a, 2003) (Figure 4.6). The APXS bombards a surface sample with alpha particles and X-rays emitted from radioactive curium-244. The alpha particles are back-scattered by the sample atoms with energies diagnostic of the atom. X-rays, whose energies are again diagnostic of the sample atoms, are emitted by the sample as ionizations caused by the initial bombardment recombine, a process called X-ray fluorescence (the Viking landers also carried X-ray fluorescence spectrometers). The two techniques are combined because the alpha backscattering technique is superior at detecting lighter elements such as carbon and oxygen while the X-ray fluorescence technique is sensitive to heavier elements (elements heavier than sodium). Iron minerals are further investigated using the MER Mössbauer spectrometer, which bombards samples with gamma rays emitted from a cobalt-57 source and measures the amount of gamma-ray re-emission from the sample (Klingelhöfer *et al.*, 2003).

### 4.3.2 Crustal composition from remote sensing observations

Earth-based telescopic observations cover a limited wavelength range and are of relatively low resolution, but they provided the first insights into martian crustal

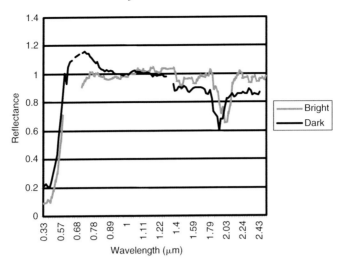

Figure 4.7 Reflectance spectra of the bright versus dark regions of Mars show important differences related to composition. These include variations in iron oxidation state, presence of pyroxenes, and existence of $H_2O$ and OH. (Data from US Geological Survey Spectral Laboratory, speclab.cr.usgs.gov.)

composition and allow long-term continuous monitoring of the planet (Singer, 1985; Bell *et al.*, 1994; Erard, 2000). Figure 4.7 shows reflectance spectra of the bright and dark regions of Mars. The reflectance increase between 0.3 and 0.75 μm is attributed to ferric iron ($Fe^{3+}$) as is the shallow absorption near 0.86 μm in the bright regions. The spectra are inconsistent with $Fe^{3+}$ occurring in crystalline form, leading to early suggestions of amorphous forms of iron oxides such as palagonite (Singer, 1985). The peak at 0.75 μm in the dark region spectrum is indicative of ferrous iron ($Fe^{2+}$), indicating differences in oxidation among the two regions. Mafic (iron- and magnesium-rich) minerals such as pyroxenes are suggested by the absorptions near 1 μm. Absorptions at 1.4, 1.9, and 3.0 μm are diagnostic of OH and $H_2O$, suggesting the presence of clays. The reddish color of the dark regions is consistent with mafic materials covered with thin alteration coatings.

Hubble Space Telescope observations provide improved spatial resolution over ground-based investigations but are limited in temporal coverage (Bell *et al.*, 1997; Noe Dobrea *et al.*, 2003). Color-ratioing of HST data reveals regional variations in composition, such as higher and/or fresher pyroxene concentrations in Syrtis Major compared to Acidalia and Utopia Planitiae (Bell *et al.*, 1997) and hemispheric differences in the distribution of hydrated minerals (Noe Dobrea *et al.*, 2003).

Orbiting spacecraft provide the highest resolution compositional information for Mars. Early spacecraft, particularly the Phobos 2 mission, gave some insights into surface compositional variations, but the first detailed mineralogic survey of the entire martian surface was performed by MGS's TES, which had a nadir surface footprint

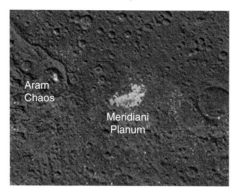

Figure 4.8 TES observations revealed a spectral signature indicative of coarse-grained crystalline hematite in the Meridiani Planum and Aram Chaos regions. The large hematite exposure in Meridiani led to the selection of this region for Opportunity's landing site. (Hematite mineralogy map courtesy of ASU/TES team.)

size of 3.15 km at the nominal orbital altitude of 378 km (Christensen *et al.*, 2001a). TES spectra are dominated by volcanic materials, particularly basalt (Christensen *et al.*, 2000a), consistent with Earth- and HST-based observations. Bright regions tend to be dust-covered, providing sparse information about the composition of the underlying bedrock (Bandfield, 2002). The dark regions comprise two distinct mineralogies with the boundary between these two surface types approximating the hemispheric dichotomy (Bandfield *et al.*, 2000). The type 1 surface unit dominates in the dark regions of the southern hemisphere and is apparently a plagioclase- and clinopyroxene-rich basalt (Bandfield *et al.*, 2000; Mustard and Cooper, 2005). Surface type 2 covers the dark regions of the northern hemisphere and has been interpreted as unaltered basaltic andesite or andesite (Banfield *et al.*, 2000) or a weathered basalt (Wyatt and McSween, 2002; McSween *et al.*, 2003). TES revealed outcrops of crystalline gray hematite ($a$-$Fe_2O_3$) in Terra Meridiani, Aram Chaos, and scattered through Valles Marineris (Figure 4.8), which were interpreted as a chemical precipitate from Fe-enriched aqueous fluids (Christensen *et al.*, 2001b). TES also detected regional concentrations of olivine (Hoefen *et al.*, 2003) and orthopyroxenite (Hamilton *et al.*, 2003). Carbonates, which might be expected from interactions of the $CO_2$-rich atmosphere with past surface water, were not detected nor were sulfates identified (Christensen *et al.*, 2001a; Bandfield, 2002).

MO THEMIS and MEx OMEGA have expanded upon the TES results with higher spectral and spatial resolutions (100 m for THEMIS, 0.3–5 km for OMEGA) and find greater mineralogic diversity across the surface (Bibring *et al.*, 2005, 2006; Christensen *et al.*, 2005). Olivine-rich basalts occur in widely separated regions, including Nili Patera, Ganges Chasma, and Ares Vallis (Christensen *et al.*, 2005; Mustard *et al.*, 2005) and outcrops of olivine are associated with some impact

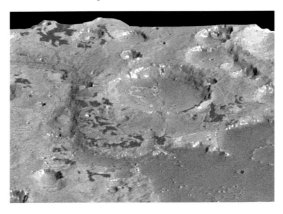

Figure 4.9 OMEGA observations have revealed localized outcrops of a variety of minerals. This view of Marwth Vallis shows outcrops (dark regions) of hydrated minerals. (Image SEMDVZULWFE, ESA/OMEGA/HRSC.)

craters and basins (Mustard *et al.*, 2005). Quartz- and plagioclase-rich granitoid rocks are exposed in the floors and central peaks of a few impact craters (Bandfield *et al.*, 2004). While much of the martian surface is composed of basalt, highly evolved high-silica (dacite) lavas have been identified in small regions such as Nili Patera in Syrtis Major (Christensen *et al.*, 2005). High-calcium clinopyroxene dominates in low-albedo volcanic regions, dark sand, and crater ejecta while low-calcium orthopyr-oxene is found in moderate to bright outcrops within the ancient terrain (Mustard *et al.*, 2005). Hydrated minerals (clays), particularly phyllosilicates, are exposed in many of the older terrain units (Bibring *et al.*, 2005; Poulet *et al.*, 2005) (Figure 4.9). Carbonates continue to avoid detection (Bibring *et al.*, 2005), but hydrated sulfates have been detected in the north polar region (Langevin *et al.*, 2005a) and within layered deposits of probable sedimentary origin (Gendrin *et al.*, 2005).

Bibring *et al.* (2006) noted an age correlation for specific mineralogies and suggested a mineralogy-based evolutionary sequence for Mars. Their earliest period (corresponding to early to middle Noachian (Section 5.2)) experienced large amounts of aqueous alteration, producing the phyllosilicates observed in the ancient terrains. Bibring *et al.* (2006) call this the "phyllosian period," in reference to the phyllosilicates. The phyllosian period was followed by the "theiikian period," characterized by acidic aqueous alteration processes which produced the sulfur deposits found in localized regions of Mars. Bibring *et al.* (2006) propose that this evolutionary change was produced by increased volcanic activity near the end of the Noachian period. The theiikian period extended from the late Noachian through the early Hesperian, when it transitioned into the "siderikian period." This latter period extends to the present day and is characterized by lack of liquid water on the global scale and for long time periods, as indicated by the presence of ferric oxides.

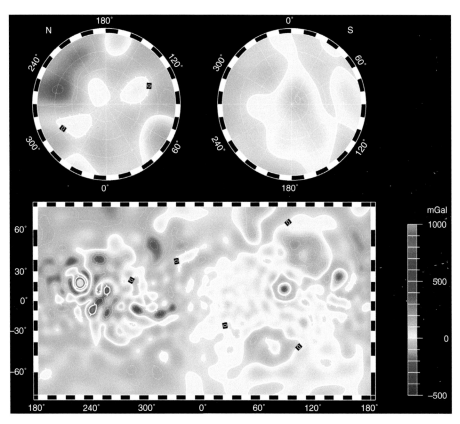

Figure 3.5 Gravity anomaly map, as measured from the vertical accelerations of MGS. A strong positive gravity anomaly is seen between 240°E and 300°E longitude which corresponds to the Tharsis volcanic province, while a strong negative anomaly in the equatorial region between 300°E and 0°E correlates with Valles Marineris. However, many topographic features do not correlate with gravity anomalies. (Image PIA02054, NASA/Goddard Space Flight Center (GSFC).)

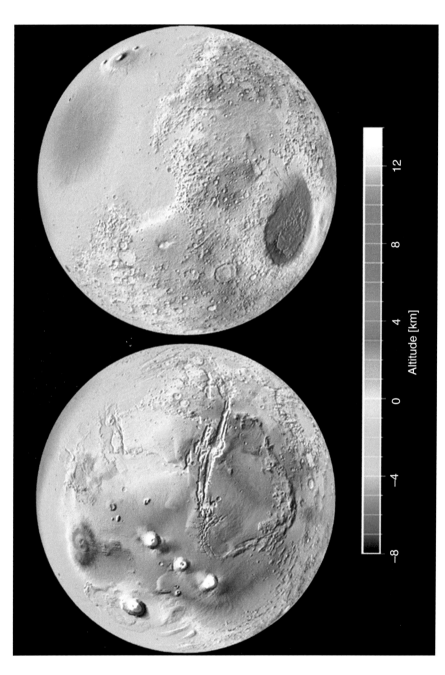

Figure 3.6 MOLA revealed the large variations in topography visible across the martian surface. Topography ranges from a high at the summit of Olympus Mons to a low in the bottom of the Hellas Basin. (Image PIA02820, NASA/JPL/GSFC/MGS MOLA team.)

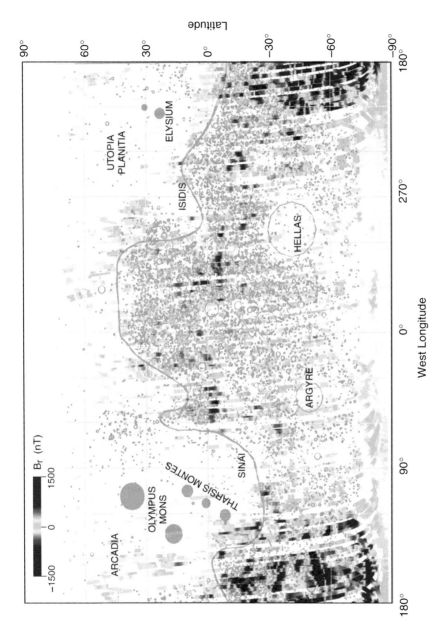

Figure 3.8 MGS's MAG/ER experiment revealed remnant magnetization within ancient rocks across the martian surface. The areas of the strongest magnetizations (red/blue colors) correspond to ancient rocks in the Terra Cimmeria and Terra Sirenum regions. Open circles indicate impact craters and filled circles are volcanoes. The dichotomy boundary is indicated by the solid line. (Image PIA02059, NASA/GSFC/MGS MAG/ER team.)

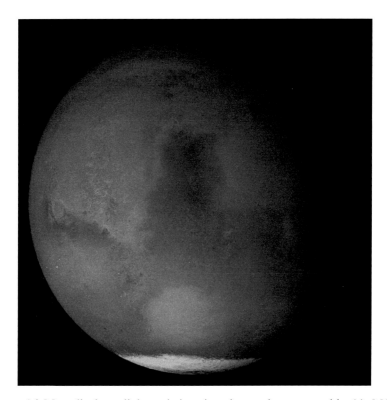

Figure 4.2 Mars displays slight variations in color, as demonstrated in this MOC image taken at $L_S = 211°$ (northern autumn/southern spring) in May 2005. The bright south polar cap is visible at the bottom of the image. The dark feature in the center is the volcanic province Syrtis Major Planum and the brighter circular feature below Syrtis Major is the Hellas impact basin. (MOC Release MOC2-1094; NASA/MSSS.)

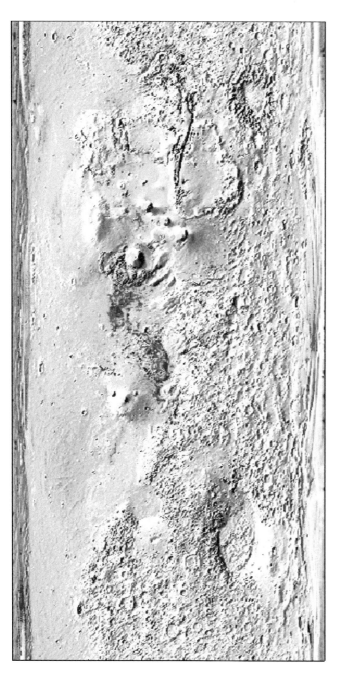
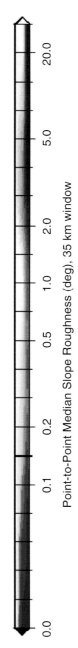

Point-to-Point Median Slope Roughness (deg), 35 km window

Figure 4.3 MOLA analysis provides estimates of the roughness of the martian surface. The southern highlands are rougher than the northern plains. (Image PIA02808, NASA/GSFC.)

Lower limit of water mass fraction on Mars

| | | | | | |
|---|---|---|---|---|---|
| 2% | 4% | 8% | 16% | 32% | >64% |

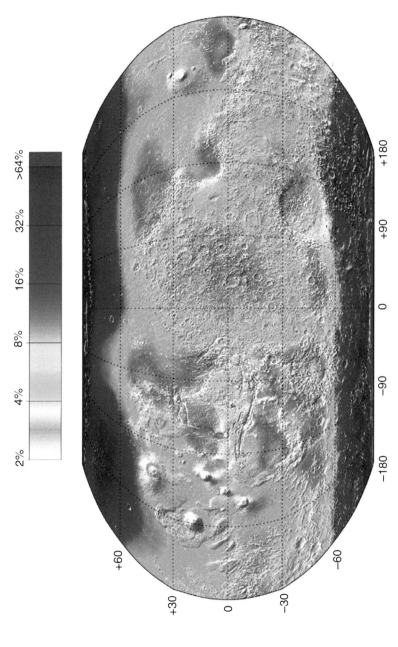

Figure 4.10 Analysis from the neutron experiments on GRS has produced information on the water mass fraction within the upper meter of the martian surface. This map shows that high water concentrations occur near the poles, which likely indicates the presence of $H_2O$ ice within the upper meter. The high $H_2O$ concentrations seen in the equatorial regions may be due to ice or hydrated minerals. (NASA/MO-GRS/Los Alamos National Laboratory.)

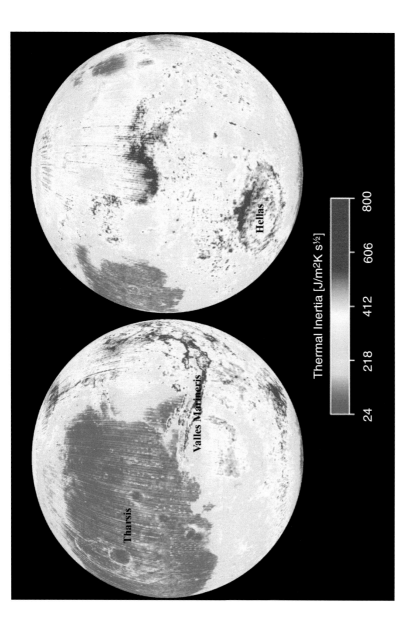

Figure 4.22 Thermal inertia provides information about particle sizes across Mars. This map, derived from TES data, shows the variation in thermal inertia across the planet. Areas of low thermal inertia, such as Tharsis, are dust-covered while regions of high thermal inertia, such as Valles Marineris and the rim of the Hellas impact basin, have high rock abundances. (Image PIA02818, NASA/JPL/ASU.)

Alteration occurring during the siderikian period has been limited to oxidation from atmospheric weathering and perhaps frost/rock interaction.

Information on the elemental concentrations of the martian crust comes from GRS analysis. Figure 4.10 shows the distribution of $H_2O$, as indicated by the detection of H, within the upper meter of the surface based on thermal, epithermal, and fast neutron analysis (Feldman *et al.*, 2004a). Regions poleward of $\sim\pm50°$ display high amounts of $H_2O$, ranging in concentration from 20% to 100% water equivalent by mass. Arabia and the highlands region south of Tharsis and Elyisum show higher $H_2O$ concentrations than the rest of the highlands, with water-equivalent concentrations of 2% to 10% by mass. While the high-latitude reservoirs likely contain substantial concentrations of $H_2O$ ice, the high-H equatorial regions could consist of either buried $H_2O$ ice or hydrated minerals.

GRS also has mapped the distribution of Cl, Fe, K, Si, and Th across the martian surface (Taylor *et al.*, 2006). Chlorine is highly mobile and can be affected by aqueous processes. High concentrations of Cl are found covering much of Tharsis, perhaps emplaced by volcanic outgassing. Older volcanic regions in eastern Tharsis and Syrtis Major and younger volcanism in Elysium show moderate Cl, Fe, and H concentrations but low K, Si, and Th. Arabia Terra is unique among the highlands regions, showing higher Th concentrations in addition to its high H signature. Most of the southern highlands have higher Si and Th concentrations than Cl, Fe, H, and K. The northern plains display high Si, Fe, K, and Th. The correlation between elemental concentrations and geologic units suggests that the elemental analysis will provide important insights into the evolutionary processes shaping the martian surface.

### 4.3.3 *Crustal composition from martian meteorite analysis*

The martian meteorites are geochemically subdivided into an orthopyroxenite (ALH84001), basaltic shergottites, olivine–phyric shergottites, lherzolitic shergottites (plagioclase-bearing peridotites), olivine-rich nakhlites, and dunites (chassignites) (Section 2.2.3) (McSween *et al.*, 2003). ALH84001 samples the early martian cumulate crust, but the other martian meteorites originated from young volcanic regions. All of the martian meteorites have higher concentrations of pyroxene than plagioclase, opposite of the TES results for the dark regions of Mars, which led McSween (2002) to argue that the SNCs originated from the dust-covered young volcanic regions of Tharsis and/or Elysium. If this is correct, the SNCs represent a biased sampling of a small region of Mars, and extrapolation of SNC crustal results to the entire planet is unjustified.

The compositional differences among the shergottites have been attributed to contamination of the magma through inclusion of early crustal material (Borg *et al.*, 1997). Alteration products resulting from fluid–rock interactions have been detected

Lower limit of water mass fraction on Mars

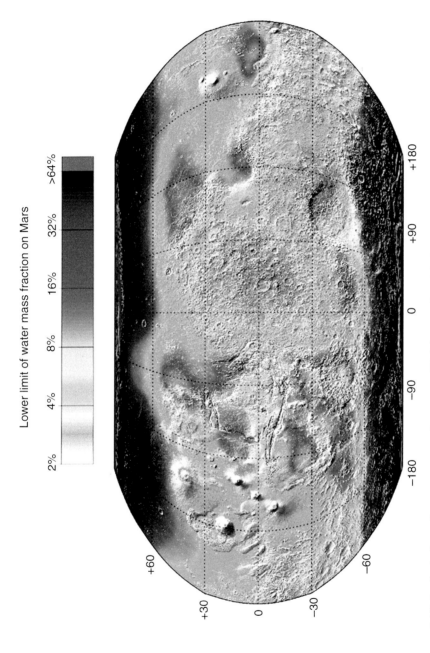

Figure 4.10 Analysis from the neutron experiments on GRS has produced information on the water mass fraction within the upper meter of the martian surface. This map shows that high water concentrations occur near the poles, which likely indicates the presence of $H_2O$ ice within the upper meter. The high $H_2O$ concentrations seen in the equatorial regions may be due to ice or hydrated minerals. (NASA/MO-GRS/Los Alamos National Laboratory.) See also color plate.

Figure 4.11 The round objects with thin rims are carbonate globules found in ALH84001. The globules are about 200 μm in diameter and consist of calcium- and iron-rich carbonates with magnesium carbonate rims. (Image courtesy of Monica Grady, British Museum of Natural History.)

in several of the martian meteorites. ALH84001 contains large quantities of carbonates (Figure 4.11), which were deposited in rock fractures either by evaporation of brines (Warren, 1998) or through reactions with hydrothermal fluids (Romanek *et al.*, 1994). Alteration products from rock–water interactions are found in several of the martian meteorites (Treiman *et al.*, 1993; Bridges and Grady, 2000). These results indicate that the crustal regions sampled by the martian meteorites are primarily composed of volcanic materials which have interacted with groundwater and/or surface water throughout their histories.

### 4.3.4 Crustal composition from in situ analysis

*In situ* analyses can provide strong constraints on the elemental and mineralogical compositions of the specific landing sites. Five missions have returned compositional information from the surface of Mars: Viking 1 lander (VL1) in Chryse Planitia, Viking 2 lander (VL2) in Utopia Planitia, Mars Pathfinder (MPF) in the outwash region of Ares Valles, MER Spirit rover in Gusev Crater, and MER Opportunity rover in Meridiani Planum. VL1, VL2, and MPF landed within the younger, lower-elevation northern plains while Spirit and Opportunity are exploring older regions of the planet. VL1 and VL2 analyzed the soil within reach of their robotic arms using X-ray fluorescence spectrometers (Clark *et al.*, 1977, 1982). Those results suggested that the soil contained higher concentrations of Fe, S, and Cl than terrestrial continental soils, but lower A1, suggesting creation from mafic to ultramafic rocks. Soil composition was amazingly similar despite the 4500-km separation of the two landers, likely due to deposition of airborne dust. Dust sampled by MPF, Spirit, and Opportunity displays a similar composition to that

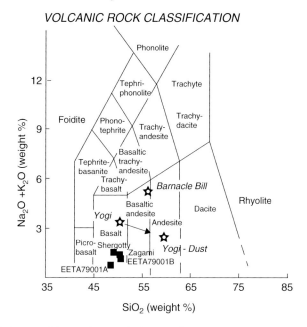

Figure 4.12 Compositions of several of the rocks and dust analyzed by MPF (stars) and some of the martian meteorites (squares) are shown on this diagram of volcanic rock classifications. The rocks analyzed by MPF tend to show higher andesitic compositions compared to the martian meteorites which tend to be basaltic. (NASA/JPL/MPF.)

observed at VL1 and VL2, suggesting that global dust storms homogenize the soil across the planet (Rieder *et al.*, 1997b; Yen *et al.*, 2005).

MPF was able to identify elements that were below VL1 and VL2 detection limits through use of the APXS and was the first mission to analyze rocks. The rocks were covered with $Fe^{3+}$ coatings of dust and some weathering rinds, making it difficult to obtain actual rock compositions (McSween *et al.*, 1999; Morris *et al.*, 2000). The MPF rocks appear to be of andesitic composition, formed by fractionation of tholeiitic basaltic magmas during early melting of the martian mantle (Figure 4.12) (McSween *et al.*, 1999). Soil composition is not an exact fit to the adjacent rocks, leading to the suggestion that the soils resulted from hydrous alteration of basalt mixed with material derived from the andesitic rocks at the landing site (Bell *et al.*, 2000).

Spirit and Opportunity have provided the best mineralogic analysis of surface materials to date. The combination of the Mini-TES, APXS, and Mössbauer instruments together with the Rock Abrasion Tool (RAT) to grind off surface alteration rinds and the Microscopic Imager (MI) to investigate rock and soil textures has led to significant insights into the geologic evolution of these two sites.

Gusev Crater was selected as the Spirit landing site based on geomorphic analysis suggesting that its floor was covered by paleolake sediments (Golombek *et al.*, 2003).

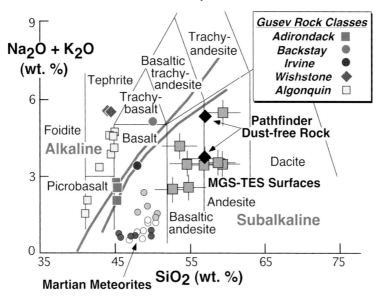

Figure 4.13 Rocks investigated by Spirit in Gusev Crater display a wide range of volcanic compositions ranging from basalts to andesites. Rocks analyzed by MPF and remote sensing data from MGS TES are included for comparison. (NASA/ JPL-Caltech/University of Tennessee.)

Spirit operated on the plains surrounding its landing site for the first 157 sols of its mission. The rocks and soil on the Gusev plains were discovered to not be of sedimentary origin but are derived from olivine-rich basaltic lava flows (Christensen *et al.*, 2004b; Gellert *et al.*, 2004, 2006; McSween *et al.*, 2004, 2006; Morris *et al.*, 2004, 2006). The composition is similar to that of the olivine-phyric shergottites, but is distinct from that of the basaltic shergottites and the andesite-rich rocks studied at the MPF landing site (Figure 4.13). This suggests that the source magmas of the Gusev plains originated from great depth in the mantle without undergoing subsequent fractionation (McSween *et al.*, 2006). Starting on sol 157 of the mission, Spirit left the volcanic plains and climbed into the elevated region called Columbia Hills (Figure 4.14). Rocks of Columbia Hills include basalts, ultramafic sedimentary rocks cemented with sulfates, and clastic rocks of varying composition (Squyres *et al.*, 2006). The Columbia Hills rocks show varying degrees of aqueous alteration and many of the rocks appear to be altered ejecta deposits from impact craters. Spirit's wheels revealed that bright material with high concentrations of salts lies under the red dust (Figure 4.15). Squyres *et al.* (2006) propose that the earliest history of these deposits was dominated by impact events and substantial amounts of water. Flooding in Ma'adim Vallis (which empties into Gusev Crater) transported

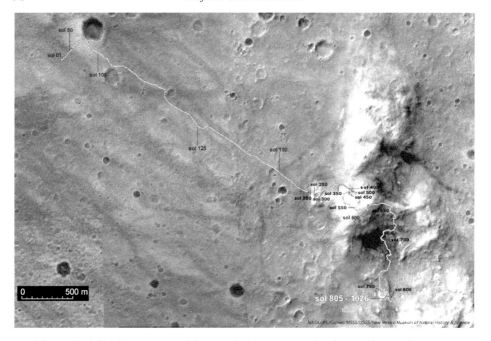

Figure 4.14 Spirit has traversed from its landing site on the floor of Gusev Crater up to Columbia Hills. This image shows the traverse path of the rover from its landing on 3 January 2004, through its climb in Columbia Hills, to its position on 22 November 2006. (NASA/JPL/Cornell/MSSS/USGS/New Mexico Museum of Natural History and Science.)

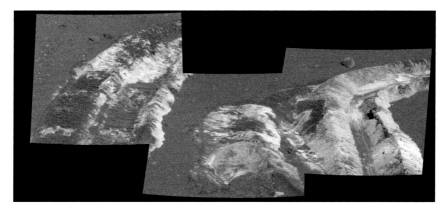

Figure 4.15 Spirit's wheels dig trenches into the fine-grained martian soil as it moves. This image looks back at the rover tracks within Columbia Hills as the rover headed for McCool Hill. The trenches reveal bright material underlying the darker surface dust. Compositional analyses of bright material exposed in several places along Spirit's traverse indicate these deposits contain salts, suggestive of water activity within this region in the past. (NASA/JPL-Caltech/Cornell University.)

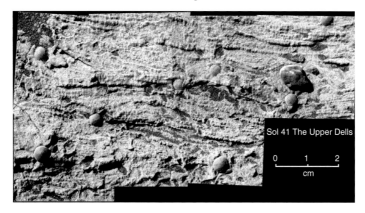

Figure 4.16 Opportunity's MI imaged hematite spherules ("blueberries") and non-horizontal layering in the Upper Dells segment of Eagle Crater's wall. The layering is suggestive of ripple patterns produced by moving water. The hematite spherules likely formed from precipitation in iron-rich water. (NASA/JPL/Cornell University/USGS.)

sand into this area which subsequently formed the observed sandstones. The water eventually evaporated, the materials were uplifted prior to the basaltic flooding which covered the surrounding plains, and the Columbia Hills region has undergone limited geologic activity during the cold, dry conditions dominating in recent times.

Opportunity landed in Meridiani Planum, the largest outcrop of gray crystalline hematite (alpha-$Fe_2O_3$) detected by TES (Golombek *et al.*, 2003). Almost immediately the rover discovered evidence that water had existed in this region (Squyres *et al.*, 2004c). The walls of 20-m-diameter Eagle Crater (in which Opportunity landed) displayed cross-stratification and ripple patterns indicative of sedimentary deposition (Figure 4.16). The unique compositional signature of this region is caused by hematite-rich spherules (nicknamed "blueberries" because they are found embedded in the sedimentary rocks like blueberries in a muffin) found in the rocks and strewn over the surface. On Earth, hematite concretions, similar to the Meridini spherules, are precipitated from Fe-rich fluids mixing with oxidizing groundwater (Chan *et al.*, 2004) and a similar process may occur on Mars (Morris *et al.*, 2005). Empty depressions within the rock, called vugs, are probably molds of salt crystals which formed in the rock and subsequently fell out or dissolved (Figure 4.17) (Herkenhoff *et al.*, 2004). Compositional analysis discovered that most of the fine-grained materials in this area were derived from basalt. The rocks contain high concentrations of various salts (Cl, Br, S) (Figure 4.18) (Rieder *et al.*, 2004) and sulfur minerals including jarosite ($NaFe_3(SO_4)_2(OH)_6$) (Klingelhöfer *et al.*, 2004). Opportunity discovered similar mineralogies at larger craters as it traversed Meridiani Planum (Figure 4.19), leading scientists to the conclusion that Meridiani Planum hosted a salt–water acidic sea in the past. Acid–sulfate weathering in

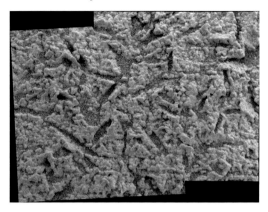

Figure 4.17 Opportunity's MI instrument revealed empty depressions within a rock called El Capitan. These depressions, called vugs, are casts of salt crystals which originally formed within the rock and subsequently dissolved or fell out. (NASA/ JPL/USGS.)

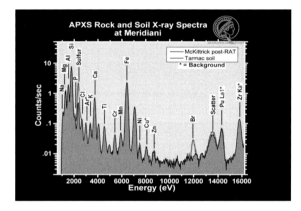

Figure 4.18 APXS analysis of the rocks exposed in Eagle Crater reveals high concentrations of salts and sulfur. Concentrations often increase from the surface into the rock interior, indicating the salts were incorporated into the rocks during formation rather than being produced by later weathering. These results suggest that Meridiani Planum was once covered by a salty sea. (NASA/JPL/Cornell University/Max-Planck Institute.)

an aqueous environment can explain the mineralogies encountered in Meridiani Planum (Golden *et al.*, 2005).

### 4.3.5 Summary of martian crustal composition

Remote sensing, martian meteorite analysis, and *in situ* investigations by the landed missions all suggest that Fe-rich volcanic materials dominate on Mars. Compositions of the dust-covered bright regions are difficult to constrain, but the presence of ferric oxides suggests that much of this material is weathered Fe-rich

Figure 4.19 Opportunity descended part way into Endurance Crater and investigated the layers exposed along a wall called Burns Cliff. Endurance Crater is larger and deeper than Eagle Crater where Opportunity landed and shows that the saltwater sea that produced the features at Eagle Crater existed for a longer period of time. (NASA/JPL.)

volcanic materials. The low-albedo regions can be subdivided into basalt (southern highlands) and andesite (northern plains). Andesitic materials could be either evolved magmas resulting from fractionation within basaltic magma chambers, or the result of weathering of basaltic materials. Water was abundant early in the planet's history, as evidenced by the presence of phyllosilicates in the old terrains and surface evidence from the Spirit and Opportunity rovers. Intense volcanism interacting with surface water near the end of the Noachian period may have led to the formation of the sulfate deposits detected by TES, THEMIS, OMEGA, and Opportunity. Acid–sulfate weathering has dominated at least in Meridiani Planum and perhaps over most of Mars. This can explain the lack of carbonates detected on the surface since carbonates easily disintegrate in acidic weathering conditions. For the last ~3 Ga, martian weathering has been dominated by dry processes, with only localized regions of surface water activity, as indicated by the presence of minerals like olivine which is rapidly altered by aqueous processes. Dust mixed into the soil at all five landing sites is similar in composition, probably because of the homogenizing effect of global dust storms.

## 4.4 Physical characteristics of surface materials

Mars' color and albedo variations are dictated not only by composition but also by grain size. Terrestrial materials are classified based on size, ranging from dust to rocks (Table 4.2). Martian materials are subdivided into rocks (pebbles, cobbles, and larger material), drift material (fine-grained and cohesive material, primarily composed of clay-sized particles), crusty-to-cloddy material (clay-sized grains weakly cemented by salts), and blocky material (composed of sand-sized and

Table 4.2 *Grade scale for small particles*

| Diameter (mm) | Particle |
| --- | --- |
| <0.004 | Clay |
| 0.0004–0.00625 | Silt |
| 0.00625–0.125 | Very fine sand |
| 0.125–0.25 | Fine sand |
| 0.25–0.5 | Medium sand |
| 0.5–1.0 | Coarse sand |
| 1.0–2.0 | Very coarse sand |
| 2.0–4.0 | Granule |
| 4.0–64 | Pebble |
| 64–256 | Cobble |

smaller grains which are strongly cemented by salts) (Moore and Jakosky, 1989). The sizes of surface materials at a specific location depend on the geologic processes which have operated in that region. Bedrock is broken into rocks that are weathered into the smaller materials.

Weathering processes are divided into physical and chemical weathering. Physical weathering is the mechanical breakage of larger materials into smaller pieces. For example, when water seeps into cracks in a rock and then expands upon freezing, the forces it exerts on the rock during expansion can eventually fracture the rock into smaller pieces. Chemical weathering is the alteration of one mineral into another, which often weakens the chemical bonds holding the rock together. Chemical weathering also includes the dissolution of minerals – thus, the vugs seen in rocks at the Opportunity landing site indicate that dissolution of salt crystals, and thus chemical weathering, has occurred.

### 4.4.1 Regolith

The uppermost layer of the surface, composed of the fragments produced by weathering processes, is called the regolith. The terms "regolith" and "soil" are often used interchangeably in planetary applications, although terrestrial geologists argue that biologic activity is an important component of the composition and mixing processes of "soil." Fragment/clast size is expected to generally increase with depth. The martian regolith likely contains mixtures of soil and ice, particularly at the higher latitudes. A possible cross-section of the martian regolith is shown in Figure 4.20.

The color of the martian regolith ranges from bright red dust to darker red and gray material. The differences in color are primarily due to variations in iron mineralogy, degree of alteration, and particle sizes and shapes. The darkest landing site yet

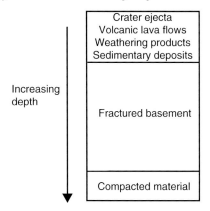

Figure 4.20 Idealized cross-section of the martian crust. The upper level is the regolith, composed of fragmented material from ejecta, volcanic flows, sedimentary deposits, and weathering products. The middle layer is basement, fractured by geologic processes such as impact cratering. At some depth, the overlying pressure is great enough to cause self-compaction of the material. (After Clifford, 1993.)

investigated is Meridiani Planum (albedo ~0.12), which is largely covered by fine-grained sand (≤150 μm) of Fc-rich basaltic composition and hematite spherules (Figure 4.21). Meridiani Planum is less dusty than the other landing sites, probably because of eolian deflation which leaves a dark cohesionless lag deposit on the surface (Soderblom *et al.*, 2004). This dark surface layer is underlain by higher-albedo soils in many locations, as revealed by trenches excavated by the rover's wheels (Figure 4.15). The crests of small dunes and drifts encountered by Opportunity across the Meridiani plains are armored by millimeter-sized rounded granules.

In contrast, the regolith at the Spirit landing site in Gusev Crater generally consists of five components (Greeley *et al.*, 2006). The topmost layer is a thin deposit (<1 mm) of dust which has settled out of the atmosphere. This is underlain by a lag deposit of coarse sand and granules, under which is a layer of subangular fragments larger than a few millimeters in size. A cohesive crust ("duricrust") several millimeters thick is next, with a layer of dark soil forming the bottom of the regolith. Gusev Crater regolith displays many of the same general characteristics as regolith at the MPF (Moore *et al.*, 1999) and Viking landing sites (Moore and Jakosky, 1989), although the VL1 and MPF sites display more fine-grained drifts than either VL2 or Gusev. Martian soil-like deposits are similar to moderately dense soils on Earth. Table 4.3 lists some of the mechanical properties of soils at the three rover sites.

All five landers have carried experiments to investigate the magnetic properties of the martian regolith. The Viking landers carried two magnets that were attached to the sampling arm so they could be directly immersed into the soil. The magnets had magnetic field strengths of 0.25 tesla (T) and 0.07 T. A third magnet with a field

Table 4.3 *Soil properties at the three rover sites*

|                      | MPF[a]          | Spirit[b]                    | Opportunity[c]              |
|----------------------|-----------------|------------------------------|-----------------------------|
| Friction angle       | 30°–40°         | ~20°                         | ~20°                        |
| Soil bearing strength |                | 5–200 kPa                    | ~80 kPa                     |
| Cohesion             | 0–0.42 kPa      | ~1–15 kPa                    | ~1–5 kPa                    |
| Angle of repose      | 32.4°–38.3°     | up to 65°                    | >30°                        |
| Soil bulk density    | 1285–1581 kg m$^{-3}$ | 1200–1500 kg m$^{-3}$  | ~1300 kg m$^{-3}$           |
| Grind energy density |                 | 11–166 J mm$^{-3}$           | 0.45–7.3 J mm$^{-3}$        |

[a] Moore *et al.* (1999).
[b] Arvidson *et al.* (2004a).
[c] Arvidson *et al.* (2004b).

Figure 4.21 Opportunity has encountered several dune and drift fields during its traverse. This navigation camera image shows dark drifts deposited by wind inside Erebus crater. (NASA/JPL-Caltech/Cornell.)

strength of 0.25 T, attached to the lander, was passively exposed to atmospheric dust. All of the magnets attracted magnetic particles – the sampler arm magnets were essentially saturated with magnetic particles after immersion in the soil (Hargraves *et al.*, 1977, 1979). The Viking analysis suggested that the particles, probably a ferromagnetic oxide, had magnetizations in the range 1–7 A m$^2$ (kg soil)$^{-1}$.

MPF carried ten magnets distributed between two arrays located on the lander. The magnetic field strengths of these magnets ranged from 0.011 to 0.280 T (Madsen *et al.*, 1999). These magnets were passive collectors of airborne dust and the results of this experiment were similar to those for the dust at the Viking landers (Section 4.3.3). The Viking and MPF results indicated that the soil and dust on Mars must contain about 2% by weight of a ferrimagnetic mineral, suggested to be either maghemite (gamma-$Fe_2O_3$) or magnetite ($Fe_3O_4$) with maghemite being the preferred component.

Spirit and Opportunity each carry seven magnets, four of which are integrated into the RAT, two on the front of the rover near the Pancam mast, and one on the solar panels (Bertelsen *et al.*, 2004). Magnetic particles in the soil and rocks are analyzed using the three magnets on the RAT, which have magnetic field strengths of 0.28, 0.10, and 0.07 T. The capture (0.46 T) and filter (0.2 T) magnets near Pancam and the sweep magnet (0.42 T) on the solar panels attract magnetized dust particles. Combined magnetic and Mössbauer spectroscopic analysis suggests that magnetite is the most probable magnetic carrier rather than the maghemite which was suggested from the Viking analysis (Bertelsen *et al.*, 2004).

### 4.4.2 Thermal inertia and rock abundance

Viking IRTM, MGS TES, MO THEMIS, and MER Mini-TES experiments have provided insights into the variation in grain size across the martian surface through measurement of the thermophysical characteristics of the regolith. Thermal inertia of the surface materials governs the daily thermal response of the surface to solar heating. The surface temperature due to absorption of solar radiation is the equilibrium temperature ($T_{eq}$) and can be calculated by balancing the incoming ($F_{in}$) and outgoing ($F_{out}$) radiation from the planet. $F_{in}$ is the amount of the solar flux absorbed over the planet's surface area which is exposed to the solar radiation:

$$F_{in} = (1 - A_b)\left(\frac{L_{solar}}{4\pi r^2}\right)\pi R^2 \tag{4.6}$$

where $A_b$ is the bond albedo (Eqs. 4.3 and 4.4), which measures the amount of radiation reflected by the planet's surface. Thus the amount of radiation absorbed is $(1 - A_b)$. The remainder of Eq. (4.6) is the amount of radiation received over the hemisphere in sunlight for a planet of radius $R$ and distance $r$ from the Sun. It is equivalent to Eq. (4.2). A rapidly rotating planet like Mars reradiates this absorbed energy from its entire surface; therefore the outgoing radiation emitted from the planet is

$$F_{out} = 4\pi R^2 \varepsilon \sigma T_{eq}^4. \tag{4.7}$$

The radiation emitted by a blackbody is proportional to the surface temperature of the body and is given by the Stefan–Boltzmann law:

$$F = \sigma T_{eq}^4. \tag{4.8}$$

The Stefan–Boltzmann constant $\sigma = 5.6705 \times 10^{-8}$ W m$^{-2}$ K$^{-4}$. Thus Eq. (4.7) is the product of the surface area of the planet ($4\pi R^2$) and the radiation emitted by that surface ($\sigma T_{eq}^4$). Because the planet is not a perfect blackbody, we must also include the emissivity ($\varepsilon$), which is a measure of how closely the object approximates blackbody emission ($\varepsilon$ for a blackbody $=1$). $F_{in}=F_{out}$ for a planet in thermal equilibrium, giving us the relationship for $T_{eq}$:

$$T_{eq} = (1 - A_b)\left(\frac{L_{solar}}{16\pi r^2 \varepsilon \sigma}\right). \tag{4.9}$$

Note that $T_{eq}$ depends only on the planet's heliocentric distance ($r$), $A_b$, and $\varepsilon$, not its size ($R$). The observed temperature (effective temperature, $T_{ef}$) is often higher than $T_{eq}$; in the case of Earth and Mars, this is because of greenhouse warming by the atmosphere (Section 6.1).

Daytime sunlight heats up the surface material on Mars, but that heat is reradiated during the night. Large rocks can retain this heat longer than smaller material like sand because of their larger volume compared to surface area ($V/A$). This ability for materials to retain heat (thermal inertia, $I$) is related to the thermal conductivity ($K_T$), density ($\rho$), and specific heat ($c_p$) of the material:

$$I = \sqrt{K_T \rho c_p}. \tag{4.10}$$

The value of $\rho$ can vary by about a factor of four for most geologic materials comprising terrestrial planet surfaces and $c_p$ varies by only about 10–20%. The largest changes in $I$ result from $K_T$, which can vary by up to three orders of magnitude (Christensen and Moore, 1992). Porosity, cohesiveness, and grain size are the major contributors to variations in $K_T$ and this is what dictates $I$ for a particular region. Small particles, such as dust and sand, have low $I$ values while higher $I$ values correspond to rockier locations. Thermal inertia is sensitive to depths penetrated by a subsurface thermal wave of period $P$. This depth is the thermal skin depth ($\delta$):

$$\delta \equiv \sqrt{\frac{K_T P}{\rho c_p \pi}} = \frac{I}{\rho c_p}\sqrt{\frac{P}{\pi}}. \tag{4.11}$$

Spectral and bolometric brightness temperatures derived from TES observations are used to calculate $I$ variations across the martian surface (Jakosky *et al.*, 2000; Mellon *et al.*, 2000; Putzig *et al.*, 2005). Figure 4.22 shows the global variation in $I$ calculated from nighttime observations. Values of $I$ range between 24 and 800 J m$^{-2}$ K$^{-1}$ s$^{-1/2}$ across the martian surface. Low $I$ values typically correlate with high-albedo regions, suggesting that these regions are dust-covered. Regions of higher $I$ are rockier, although the apparent high $I$ values near the north pole are an artifact resulting from measurements made near dawn when temperature variations are large. High $I$ is often correlated with low albedo, interpreted to be regions of coarse-grained sediments, rocks, and bedrock exposures. A third region consists of

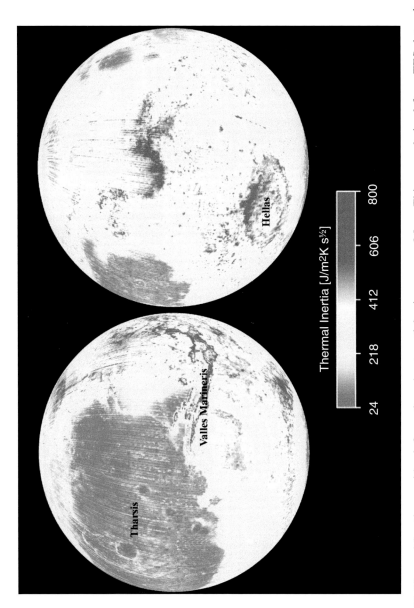

Figure 4.22 Thermal inertia provides information about particle sizes across Mars. This map, derived from TES data, shows the variation in thermal inertia across the planet. Areas of low thermal inertia, such as Tharsis, are dust-covered while regions of high thermal inertia, such as Valles Marineris and the rim of the Hellas impact basin, have high rock abundances. (Image PIA02818, NASA/JPL/ASU.) See also color plate.

99

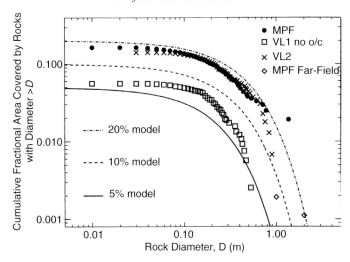

Figure 4.23 Estimated rock abundances calculated from thermal inertia considerations are compared to observed rock abundances at the VL1, VL2, and MPF landing sites. The results for MPF rock abundance are consistent with landing site predictions before launch. (Reprinted by permission of American Geophysical Union, Golombek *et al.* [1999b], Copyright 1999.)

high $I$ and intermediate albedo, interpreted as surfaces covered by duricrust interspersed with rocks and bedrock exposures (Mellon *et al.*, 2000; Putzig *et al.*, 2005).

High $I$ generally translates to a rockier surface in the $\geq$0.1–0.15-m rock size range, so that $I$ can be used to estimate rock abundances across the planet (Christensen, 1986). Rock abundance values range from 1% of the surface covered by rocks to a maximum of $\sim$35% coverage, with a modal value of 6%. Predicted rock abundance has been compared with actual values at the five landing sites and generally the two are consistent (e.g., Golombek *et al.*, 1999b, 2006) (Figure 4.23). The size–frequency distribution (SFD) of rocks follows a simple exponential curve, compatible with the distributions expected from the fracture and fragmentation of rocks (Golombek and Rapp, 1997).

### 4.4.3 Dust

Dust comprises a substantial portion of the martian soil and coats most of the rocks at the five landing sites. Winds often loft the dust into the atmosphere, transporting it around the planet. Local, regional, and global dust storms (Section 6.3.2) are the obvious manifestations of this dust transport, but the observation of dust devils (Figure 4.24) (Greeley *et al.*, 2006) and atmospheric dust (Smith *et al.*, 2000) indicates that transport of fine-grained particles occurs continuously. The dust particles are estimated to typically be <40-μm diameter and composed of ferric oxides.

Figure 4.24 Dust devils moving across the floor of Gusev Crater have been imaged several times by Spirit's Navigational Camera. This dust devil occurred on sol 486 of Spirit's mission and is about 1 km from the rover's location. (Image PIA07253, NASA/JPL.)

Adhesion of atmospherically deposited dust on the MPF and MER passive magnets indicates that the dust contains a magnetic component similar to the soils (Section 4.3.2). The global homogeneity of dust suggests a similar composition regardless of location. MPF and MER magnetic investigations suggest that airborne dust has a saturation magnetization of 2 to 4 A m$^2$ (kg soil)$^{-1}$ (Hviid *et al.*, 1997; Bertelsen *et al.*, 2004). These magnetizations are too high for maghemite particles, leading the investigators to propose magnetite as the dominant magnetic phase in the dust (Bertelsen *et al.*, 2004), a composition consistent with Mössbauer results (Morris *et al.*, 2006).

Atmospheric transport leads to electrostatic charging of the dust through collisions, particularly in dust storms and dust devils (Ferguson *et al.*, 1999; Zhai *et al.*, 2006). No discharges have been observed in martian dust devils (Krauss *et al.*, 2006), but electrostatic charging may be responsible for the accumulation of dust on rover wheels. Ferguson *et al.* (1999) estimate that the wheels on MPF's Sojourner rover acquired a charging voltage of ~60–80 V from dust buildup during its traverses. Abrasion of Sojourner's wheels suggests that martian dust has a hardness of 4.3 on the Mohs' scale of hardness, similar to platinum (Ferguson *et al.*, 1999).

# 5

# Geology

Solid bodies have their surfaces affected by geologic processes. By studying the current state of a planetary surface and applying our terrestrial experience of what features are associated with the different geologic processes, planetary geologists disentangle information about the geologic and thermal evolution of the body in question.

Geologic processes are divided into internal and external processes. Internal processes originate from within the body and include volcanism, tectonics, and mass wasting (caused by the planet's gravity). External processes originate outside of the planet's interior and include impact cratering, eolian (wind-blown), fluvial, and glacial processes. All of these processes have operated on Mars to varying extents.

## 5.1 Geologic terms and techniques

### 5.1.1 Rocks and minerals

Understanding a planet's geologic history requires development of techniques to read the record left by geologic processes. Solid bodies like Mars are composed of rocks, which are made up of minerals. A mineral is a naturally formed substance with a specific chemical composition. It can be composed entirely of one element or it can be a compound consisting of two or more elements. Minerals usually have a specific crystalline structure and changes in crystal structure, even when chemical composition remains constant, result in a different mineral.

Rocks are composed of a mass of minerals. A rock can be composed of a single mineral type or be a mixture of different minerals. Igneous rocks are rocks that solidify from molten material. The molten rock is called magma when it occurs underground and lava once it is extruded onto the planet's surface. Intrusive igneous (plutonic) rocks cool slowly underground and are characterized by large crystals. Extrusive igneous rocks form on the surface and typically cool faster than intrusive rocks, resulting in small crystals or amorphous (non-crystalline) structures.

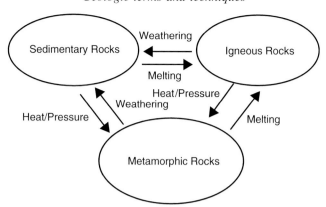

Figure 5.1 Igneous, sedimentary, and metamorphic rocks can be transformed into each other through the rock cycle.

Sedimentary rocks are composed of small fragments of other rocks, produced from weathering processes. Metamorphic rocks form when heat and/or pressure change the characteristics of an original rock without melting it. Rocks can be transformed from one type of rock into another, as demonstrated in the rock cycle (Figure 5.1).

### 5.1.2 Stratigraphic techniques

Stratigraphy is the basic principle used by geologists to interpret the geologic record. Stratigraphic analysis was first developed in 1669 by Nicolaus Steno, who realized that recent geologic events deposit their materials atop older, pre-existing layers of rock and soil. This is the principle of superposition, where the oldest layers are the most deeply buried and the younger units are near the surface (Figure 5.2). By analyzing rocks in these stratigraphic sequences, geologists can determine the types and relative timing of the geologic processes that created the layers.

Superposition is often combined with the principle of transection to constrain the relative timing of geologic events. Transection states that younger features cut older features. Transection is often seen with tectonic, fluvial, and glacial processes, which incise canyons or channels into pre-existing (and thus older) terrain. Later events can infill these features – it is not uncommon to see evidence of an ancient channel which has been filled in with later sedimentary or igneous deposits when looking at a canyon wall or a highway roadcut.

Superposition and transection are the two major stratigraphic techniques used on Earth. On Earth we have the additional advantage of being able to analyze rocks in our laboratories. Igneous rocks often contain small amounts of radioactive elements that can be used to date the time of solidification of the rock (Section 2.2.2). The age

Figure 5.2 Superposition is clearly seen in the layers exposed in Arizona's Grand Canyon. The older layers underlie younger layers. (Image by author.)

obtained by geochronologic techniques is the absolute age of the rock. If no datable rocks are available, geologists can develop a relative chronology from the stratigraphic positions of the rock layers. Absolute chronologies therefore provide actual dates for the time at which events occurred while relative chronologies only tell you whether something is older or younger than something else.

A third stratigraphic technique is commonly employed on other solid-surfaced bodies in our Solar System. Since datable samples are limited or non-existent from these bodies, planetary geologists use the number of craters as a key to determining relative chronologies and approximate absolute chronologies. Crater density is the cumulative number of craters per unit area, typically given as the number of craters greater than or equal to some diameter per $10^6 \, km^2$. The probability that a surface has experienced an impact increases with the age of that surface since impact cratering has occurred throughout Solar System history. Thus older units will have higher crater densities than younger terrains.

### 5.1.3 Crater statistical analysis

Age information from crater analysis is obtained through use of crater size–frequency distributions (SFDs) (Crater Analysis Techniques Working Group, 1979). SFDs compare the frequency of craters as a function of crater diameter. An initial assumption of SFD analysis is that such distributions approximate a power-law function for an incremental distribution:

$$N = KD^{-\alpha} \tag{5.1}$$

where $N$ is the number of craters of diameter $D$ and larger, $K$ is a constant that depends on the crater density, and $\alpha$ is the slope of the power-law function (also

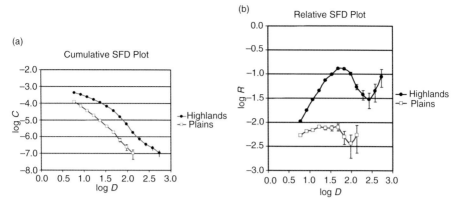

Figure 5.3 Crater size–frequency distributions (SFD) are studied using cumulative and relative plots. (a) The cumulative SFD plots for craters on the martian highlands and northern plains are shown. (b) Compare the relative SFD plots to those in (a) for the same regions. Obvious differences in the shapes of the curves are readily apparent in the relative plots.

called the population index). The cumulative SFD is an example of an incremental distribution. Figure 5.3a shows a sample cumulative plot, where the number of craters of a certain diameter and larger per unit area ($C$) is plotted against diameter on a log–log plot. Error bars are calculated from

$$\pm\sigma = C \pm \frac{C}{\sqrt{N}}. \tag{5.2}$$

SFDs for craters larger than 5 to 8 km in diameter often show an approximately straight line with a $-2$ slope, especially for younger surfaces. The negative slope indicates that smaller craters are more abundant than larger craters. The values of SFD for different surface units vary vertically on this plot depending on crater density – higher crater densities lie above lower crater densities. Thus, the relative positions of SFDs from two different terrains indicate that the surface corresponding to the higher curve is older than the surface displaying the lower curve. The cumulative plot suggests that SFDs for all terrain units can be approximated by a power-law function with a slope of approximately $-2$ at all diameters. However, this may be a reflection of the plotting technique because the cumulative nature of this technique tends to smooth out frequency variations within a particular diameter range.

The relative or *R*-plot technique overcomes the smoothing problem of cumulative plots. The *R*-plot is a differential plot of the form (Crater Analysis Techniques Working Group, 1979)

$$dN = CD^{-\beta}dD \tag{5.3}$$

where d$N$ is the number of craters with diameters in the range d$D$, $D$ is the mean diameter of the range, $C$ is a constant, and $\beta$ is the differential population index. Because this equation is derived by differentiating the incremental form given in Eq. (5.1), $\beta = \alpha + 1$. The $R$-plot differs from the cumulative plot in two ways: (1) it only includes the number of craters within a particular size range, thus frequency variations are not reduced as with the cumulative technique, and (2) it normalizes the curves to a common slope index of $-3$ (corresponding to the cumulative slope index $\alpha = 2$). The sizes of the diameter ranges (or bins), d$D$, are set at $\sqrt{2}$ increments; hence d$D$ may be 2 km to 2 $\sqrt{2}$ km ($= 2.8$ km), the next bin will be from 2.8 km to $2.8\sqrt{2}$ km ($= 4.0$ km), etc. The relative technique plots the geometric mean diameter of the bin on the horizontal axis and the normalized parameter $R$ on the vertical axis of a log–log plot. $R$ is given by

$$R = \frac{\overline{D}^3 N}{A(D_b - D_a)} \qquad (5.4)$$

where $N$ is the number of craters with diameters in the range $D_a$ (lower diameter limit of bin) to $D_b$ (upper diameter limit, $D_b = D_a\sqrt{2}$), $\overline{D}$ is the geometric mean of the diameter bin ($\overline{D} = \sqrt{D_a D_b}$), and $A$ is the area over which the craters are counted. SFDs following the $-2$ slope suggested from cumulative analysis will plot as a horizontal line on the $R$-plot, while SFDs following other slopes will plot as inclined lines. Error bars are calculated as

$$\pm\sigma = R \pm \left(\frac{R}{\sqrt{N}}\right). \qquad (5.5)$$

Figure 5.3b shows the $R$-plot of the same terrain units displayed in the adjacent cumulative plot. The lower curve (i.e., the younger terrain) approximates a horizontal line on the $R$-plot, corresponding to a power-law function of $-2$ cumulative slope. However, a single-sloped power-law function does not approximate the upper (older) curve. The upper curve is representative of the SFDs seen on heavily cratered regions of the Moon, Mercury, and Mars, while the lower curve is similar to that seen on the lunar maria and martian northern plains.

A third type of crater plot combines properties of the cumulative and relative plots. This is a log-incremental SFD, primarily used by W. Hartmann and colleagues (e.g., Hartmann, 2005). In this technique, the number of craters within a particular diameter bin per unit area is plotted on the ordinate while the diameter is plotted on the abscissa (Figure 5.4). This plot is similar to the $R$-plot except the results are not normalized to the $-3$ differential slope.

The cause of the multisloped highlands SFD curve is still debated. One suggestion is that this curve represents the SFD of terrains saturated with craters

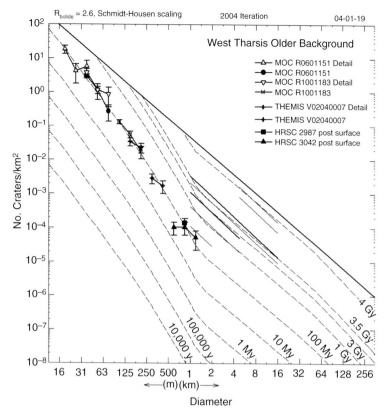

Figure 5.4 Isochron plots provide insights into the absolute ages of surface units based on crater density. This plot of craters in a region west of Tharsis indicates that the surface is approximately 100 Ma old. (Image courtesy of William Hartmann, Planetary Science Institute.)

(Hartmann, 1984, 1997). Saturation refers to surfaces which are covered with so many craters that any new impact erases an existing one, keeping the crater density approximately constant. No age information (other than "very old") can be obtained from saturated surfaces. While ancient surfaces may display crater saturation in some diameter ranges (particularly at the smaller crater diameters), other diameter ranges appear to be largely unaffected. Observational evidence against saturation is seen in places such as on the floor of the Moon's Orientale Basin and on the martian ridged plains. These terrains have lower crater density than the heavily cratered regions, but still display the multisloped curve shape (Barlow, 1988; Strom *et al.*, 1992).

A second possible explanation for the multisloped curve of the highlands is that erosion has preferentially destroyed the smaller craters, causing a change in slope at the lower-crater-diameter end. While erosion definitely affects the SFDs, it does not appear to be the total explanation for the frequency downturn in the 5 to ~70 km diameter range. Heavily cratered regions on the Moon, Mercury, and Mars all show the same multisloped curve, but the three bodies have experienced very different obliteration histories. In addition, the SFDs of relatively fresh impact craters (those still retaining ejecta blankets; by definition, fresh craters have not experienced much erosion) on heavily cratered martian terrains also show the downturn at smaller diameters (Barlow, 1990).

The different SFDs could also indicate two crater-production populations with different size–frequency distributions. The multisloped curve is seen on surfaces believed to date from the Late Heavy Bombardment (LHB) period (Section 2.1.2) while the flatter curve is seen on younger terrains. The current impactor population in the inner Solar System is derived primarily from asteroids with some contribution from comets (Bottke *et al.*, 2002) and the crater SFD is consistent with the asteroid SFD (Bottke *et al.*, 2005). LHB impactors likely originated from the planetesimal disk disrupted by the migration of the outer planets (Gomes *et al.*, 2005), although a source in the main asteroid belt also has been proposed (Strom *et al.*, 2005). If the SFDs reflect two different impactor populations, they imply that the LHB population was enriched in mid-sized and large impactors and depleted in smaller objects compared to the present-day asteroid SFD. The theory of different impactor populations indicates that not only can the SFD curves tell us the crater densities and thus the relative ages of the different terrain units but also whether the craters formed during the LHB or post-Heavy Bombardment periods. Such an analysis applied to Mars suggests that ~60% of the planet's surface dates from LHB while the remaining 40% formed more recently (Barlow, 1988).

Crater statistical techniques provide the best chronologic information when the number of craters per unit area is greatest, thereby reducing the uncertainties. For craters larger than ~5-km diameter, this requires performing the crater analysis over relatively large areas to obtain statistically viable results. Geologic units are seldom uniform over such large areas, resulting in the age information derived from such analyses being averages of several incorporated units. Thus, planetary geologists prefer using craters <5-km diameter, which are more abundant than the larger craters and allow statistically significant results to be obtained for regions of small areal extent (e.g., Hartmann *et al.*, 1999; Neukum *et al.*, 2004).

However, chronologic information derived from small-crater counts is affected by two major processes. Erosion, as noted previously, preferentially destroys smaller craters, leading to younger age estimates on surfaces where erosion has dominated. The other problem is secondary crater contamination, which may be responsible for a steepening of the SFD at ≤1-km diameter. Secondary craters are small craters formed from material ejected during formation of a larger (primary) crater. Secondary clusters and those close to the primary crater ("obvious secondaries") can be easily identified and avoided, but distant secondaries can be difficult to distinguish from small primary craters. Inclusion of secondaries produces a higher crater density and thus indicates an older age for the surface. Some researchers argue that secondaries are a minor contributor to the SFD since fragmentation and grinding of ejected debris eliminates material in the smallest diameter ranges (Werner *et al.*, 2006). Hartmann (2005) argues that the SFD is a combination of primary and distant secondaries which approximates the primary production curve over time. Clustering analysis suggests that the majority of small craters on Europa are secondaries, a result which is probably applicable to other Solar System bodies (Bierhaus *et al.*, 2005). The 10-km-diameter martian crater Zunil displays thermally distinct secondary crater rays extending over 300 crater radii away from the primary crater (McEwen *et al.*, 2005). The large number and wide distribution of Zunil secondaries supports the assertion that most martian craters ≤1-km diameter are probable secondaries. However, secondary crater production is enhanced on young lava plains and occurs only in association with larger craters in regions with thicker regolith (Hartmann and Barlow, 2006). This observation combined with the fact that secondary craters are concentrated in non-uniformly distributed rays indicates that secondary contamination is not evenly distributed across the entire planet. The combined problems of erosion and secondary craters lead to considerable uncertainty in chronologic information derived from small-crater analysis (McEwen and Bierhaus, 2006).

Relative ages obtained from crater statistical analysis can be used to estimate absolute ages of surface units. Radiometric dating of lunar samples returned by the Apollo and Luna missions allows correlation of crystallization ages with the crater density of the region surrounding the sample site. The resulting crater density versus age plot is the lunar crater chronology (LCC) graph (Figure 2.2). The LCC can be used to estimate absolute ages of non-sampled regions of the Moon based on the crater density of those regions.

The LCC can be extrapolated to other Solar System bodies to obtain absolute age estimates provided the following three conditions are met: (1) the population of impacting bodies is the same between the Moon and body of interest; (2) the timing of the LHB between the Moon and other body is known; and (3) the impact flux for the body relative to the Moon is known. The similarity of the SFDs for the Moon and Mars indicates that condition 1 is met. Recent numerical modeling suggesting that the LHB resulted from disruption of a planetesimal disk by migration of the outer planets (Gomes *et al.*, 2005) indicates that the beginning and end of LHB were approximately simultaneous throughout the inner Solar System, satisfying condition 2. The Mars/Moon impact rate ratio can be estimated from SFDs of the impacting objects and knowledge of impact probability for asteroids/comets crossing the orbits of the Moon and Mars (Ivanov, 2001). The ratio of the impact rate of projectiles of a certain size per unit area compared to that impact rate on the Moon is called the bolide ratio ($R_b$). The current Mars/Moon $R_b$ value is 4.93, but because of orbital eccentricity variations experienced by Mars (Section 7.4.2), the value can be as low as 2.58 (Ivanov, 2006).

The LCC can be extrapolated to Mars by utilizing the Mars/Moon impact rate ratio and scaling relationships for converting impactor size to final crater diameter (Section 5.3.1). Hartmann (2005) has produced isochron plots for Mars which allow one to estimate the formation age of a surface from crater counts (Figure 5.4). Variations in ages associated with different diameter ranges provide insights into processes such as erosion and possible secondary crater contamination.

## 5.2 Martian geologic periods

Martian history is divided into periods based on stratigraphic relationships and occurrence of certain geologic processes (Figure 5.5) (Tanaka *et al.*, 1992). The oldest period is the Noachian, named after the Noachis region in the southern highlands. Noachian terrains formed during the LHB and are thus very heavily cratered surfaces. The range of degradation associated with craters formed during this period suggests high erosion rates due to geologic processes such as rainfall and fluvial erosion (Craddock and Howard, 2002). The Noachian period is divided into early (prior to ~3.95 Ga ago), middle (between 3.95 and 3.8 Ga ago), and late (3.8– 3.7 Ga ago) periods based on crater densities (Hartmann and Neukum, 2001). The Hesperian Period (named after Hesperia Planum) represents middle martian history and is characterized by volcanic extrusions creating the ridged plains and a decline in the high impact rates that occurred during the LHB. It is subdivided into early (3.7–3.6 Ga ago) and late (~3.6–3.0 Ga ago) periods. The Amazonian Period (from Amazonis Planitia) covers the planet's most recent history during which erosion rates have been low and volcanic activity has been concentrated in the Tharsis and Elysium regions. The Amazonian Period is divided into early (~3.0–1.8 Ga ago), middle (~1.8–0.5 Ga ago), and late (~0.5 Ga ago to present) periods.

## 5.3 Geologic processes

### 5.3.1 Impact cratering

Impact craters are the most common geologic features observed on most planetary surfaces. The term *crater*, Latin for cup, was introduced by Galileo in 1610 to describe the approximately circular depressions observed on the lunar surface. Most early scientists believed that the lunar craters were of volcanic origin, primarily because of their circular appearance and terrestrial experience with volcanic craters

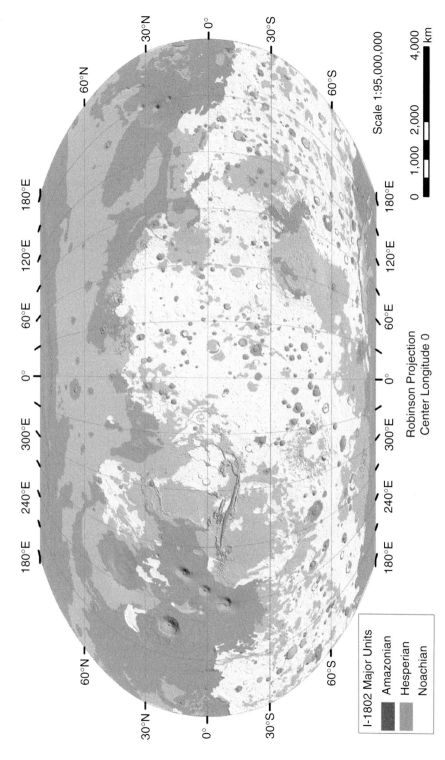

Figure 5.5 Stratigraphic techniques provide insights into the ages of surface units. This map shows the distribution of units formed during the Noachian, Hesperian, and Amazonian periods. (Image courtesy of Trent Hare, USGS.)

I-1802 Major Units

Amazonian

Hesperian

Noachian

Scale 1:95,000,000

0    1,000   2,000      4,000
                          km

Robinson Projection
Center Longitude 0

(calderas). Acceptance of the idea that craters could result from collisions of large chunks of space debris with a planetary surface was gained only after (1) laboratory experiments showed that circular craters could be produced during high-velocity (hypervelocity) impacts (Ives, 1919; Gault *et al.*, 1968), (2) nuclear and chemical explosions provided insights into the physics of impact events (cf. Roddy *et al.*, 1977), and (3) identification of shock metamorphic features in rocks surrounding Meteor Crater and Ries Crater (Shoemaker and Chao, 1962; Shoemaker, 1963).

Impact crater formation is divided into three stages (Gault *et al.*, 1968; Melosh, 1989): contact/compression, excavation, and modification. The projectile first encounters the surface during the contact/compression stage, generating shock waves which propagate through both the target and the projectile. The Hugoniot equations, derived from conservation of mass, momentum, and energy, describe the material characteristics on either side of the shock front:

$$\rho(U - u_p) = \rho_0 U \tag{5.6}$$

$$P - P_0 = \rho_0 u_p U \tag{5.7}$$

$$E - E_0 = \frac{(P + P_0)(V_0 - V)}{2} = \frac{(P + P_0)\left(\frac{1}{\rho_0} - \frac{1}{\rho}\right)}{2}. \tag{5.8}$$

The subscript 0 refers to the material properties – pressure $P$, density $\rho$, specific (per unit mass) internal energy $E$, and specific volume $V$ ($= 1/\rho$) – before passage of the shock front, while those parameters without a subscript refer to the compressed material (after passage of the shock wave) (Figure 5.6). The shock wave moves with velocity $U$ and, after the shock front has passed, the compressed material has a particle velocity of $u_p$. The reference frame when using the Hugoniot equations is the rest frame of the uncompressed material ($u_{p0} = 0$).

The shock wave pressure declines exponentially with distance from the impact site (Figure 5.7) and the shock wave turns into a seismic (elastic) wave once the pressure drops below 1 to 2 GPa. As the shock wave encounters a free surface (either the surface of the target material or the back surface of the projectile), it is reflected as a rarefaction or release wave back into the material. As rarefaction waves pass through the highly shocked material, they unload some of the pressure, resulting in melting and vaporization of some of the material. The contact and compression stage ends when the rarefaction wave engulfs and destroys the projectile. The projectile usually has traveled a distance equivalent to its diameter during the contact/compression stage, corresponding to only a few seconds or less.

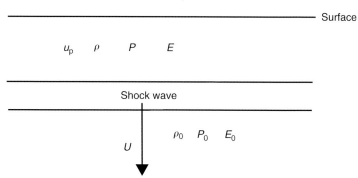

Figure 5.6 The passage of the impact-induced shock wave alters characteristics of the rock. Before the shock wave passes, the material is characterized by density ($\rho_0$), pressure ($P_0$), and specific internal energy ($E_0$). The shock wave travels with velocity $U$. Once the shock front passes, the rock acquires new values of density ($\rho$), pressure ($P$), and specific internal energy ($E$), as well as a particle velocity ($u_p$). Relationships between these before-and-after values are obtained from the Hugoniot equations.

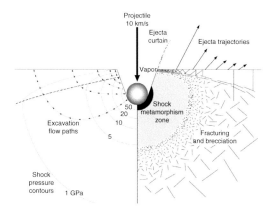

Figure 5.7 Shock-induced pressure declines with distance from the impact site. Near the impact site, pressures are high enough to induce melting and vaporization. Further from the impact, the passage of the shock wave produces fracturing of the rock. (Reprinted by permission from Lunar and Planetary Institute, French [1998], Copyright 1998.)

The excavation stage opens the crater. As the shock wave encounters the free surface, part of it is converted into kinetic energy and the rest is reflected as a rarefaction wave. The rarefaction waves will fracture the target material as long as the stress of the rarefaction wave (a tensional wave) exceeds the mechanical strength of the rock. The portion of the shock wave converted to kinetic energy will accelerate fragmented material outward. Some of this material is accelerated up and out

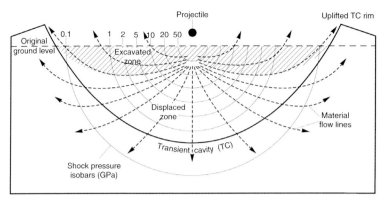

Figure 5.8 The transient cavity (TC) forms during the excavation stage, as the shock wave is converted into kinetic energy. Material can be either displaced downward, or up and out of the crater. The material ejecta outward forms the raised rim and ejecta blanket. (Reprinted by permission from Lunar and Planetary Institute, French [1998], Copyright 1998.)

of the crater, forming the ejecta blanket and part of the uplifted rim (the rest of which results from structural uplift from the shock and rarefaction waves) (Figure 5.8). The rest of the material is displaced downward, resulting in a bowl-shaped cavity called the transient crater. The transient crater has the greatest depth ($d_t$) that the crater will ever have and is between a third and a quarter of the transient crater diameter ($D_t$).

Scaling relationships are used to extrapolate results from laboratory experiments to impact events (Melosh, 1989). The value of $D_t$ is related to energy of the impact ($W$), size of the projectile ($L$), acceleration of gravity ($g$), densities of the projectile and target ($\rho_p$ and $\rho_t$, respectively), and impact angle (measured from horizontal surface) ($\theta$) through

$$D_t = 1.8\rho_p^{0.11}\rho_t^{-0.33}g^{-0.22}L^{0.13}W^{0.22}(\sin\theta)^{0.33}. \tag{5.9}$$

Smaller craters tend to be bowl-shaped depressions, approximating the shape and depth of the transient crater. These craters are called simple craters (Figure 5.9a). Depth ($d$) of a simple crater is related to its rim diameter ($D_r$) by

$$d \approx \frac{D_r}{5}. \tag{5.10}$$

Larger craters display more complicated morphologies and are called complex craters (Figure 5.9b). Complex craters are shallower compared to their size than simple craters, typically about one-tenth of the rim diameter. The diameter where craters transition from simple to complex structures is proportional to $g^{-1}$, although target characteristics also contribute. The simple-to-complex transition diameter

(a)　　　　　　　　　　　　(b)

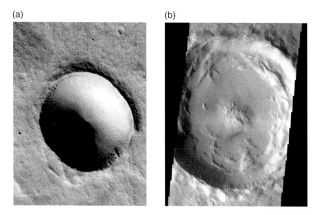

Figure 5.9 Impact craters display morphologic differences as diameter increases. (a) The smallest craters display a bowl-shaped appearance and are called simple craters. This 2-km-diameter simple crater on Mars is located near 34.4°N 241.2°E. (MOC image MOC2-1274, courtesy of NASA/MSSS.) (b) Larger craters, called complex craters, display more complicated morphologies, including central peak structures, shallower floors, and terraced walls. This complex crater is 27 km in diameter and located at 28.6°N 207.0°E. (THEMIS image V17916017, NASA/ASU.)

($D_{SC}$) on Mars is predicted to occur near 10 km, but observations place it closer to 8 km (Garvin and Frawley, 1998). The lower $D_{SC}$ probably results from substantial quantities of ice in the near-surface material reducing the strength of the target material. Variations in morphometric properties such a crater depth, volume, and rim height with latitude also are attributed to higher concentrations of subsurface ice near the poles (Garvin *et al.*, 2000a). Table 5.1 provides morphometric relationships for fresh martian impact craters.

Material ejected during crater formation forms the ejecta blanket surrounding the crater. The ejecta blanket does not contain material from the entire cavity. The depth of excavation ($d_{ex}$) for the ejected material is approximately one-third of the transient crater depth:

$$d_{ex} \approx \frac{1}{3}d_t \approx \frac{1}{10}D_t. \tag{5.11}$$

Ejecta blankets are divided into continuous and discontinuous sections. The continuous ejecta blanket abuts the crater rim and is composed of a continuous blanket of debris typically extending between one and three crater radii from the rim. The discontinuous ejecta blanket extends beyond the continuous ejecta blanket in discrete streaks radial to the crater. Secondary craters populate most of the discontinuous ejecta blanket (Figure 5.10).

Table 5.1 *Impact crater morphometric parameters as function of crater diameter (D)*

| Parameter | Simple craters | Complex craters |
|---|---|---|
| Depth ($d$) | $d = 0.21D^{0.81}$ | $d = 0.36D^{0.49}$ |
| Rim height ($h$) | $h = 0.04D^{0.31}$ | $h = 0.02D^{0.84}$ |
| Central peak height ($h_{cp}$) | — | $h_{cp} = 0.04D^{0.51}$ |
| Central peak diameter ($D_{cp}$) | — | $D_{cp} = 0.25D^{1.05}$ |
| Inner cavity wall slope ($s$) | $s = 28.40D^{-0.18}$ | $s = 23.82D^{-0.28}$ |

*Source*: Garvin *et al.* (2003).

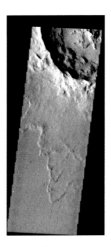

Figure 5.10 Ejecta blankets can be divided into continuous and discontinuous parts. This image of the ejecta blanket associated with a 28.3-km-diameter crater located at 23.2°N 207.8°E shows the layered continuous ejecta blanket extending outward from the rim. Beyond the edge of the layered ejecta blanket, small secondary craters are seen, which constitute the discontinuous ejecta blanket. (THEMIS image V01990003, NASA/ASU.)

Most fresh martian impact craters are surrounded by a layered ejecta blanket with one (single layer), two (double layer), or more than two (multiple layer) ejecta layers (Figure 5.11) (Barlow *et al.*, 2000). This ejecta morphology is distinct from the radial morphology seen around craters on dry bodies like the Moon. Two models have been proposed to explain this layered ejecta morphology: (1) impact into and vaporization of subsurface volatiles (Carr *et al.*, 1977; Stewart *et al.*, 2001), and (2) interaction of the ejecta curtain with the martian atmosphere (Schultz, 1992; Barnouin-Jha *et al.*, 1999a, b). While the atmosphere plays some role, most of the evidence suggests that subsurface volatiles are the dominant contributor in the formation of the layered ejecta morphologies (see review in Barlow, 2005).

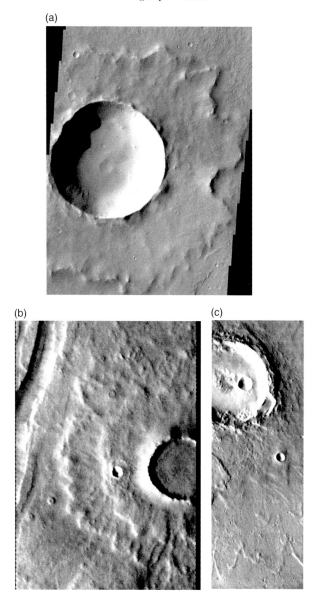

Figure 5.11 Layered ejecta blankets are divided into single layer (SLE), double layer (DLE), and multiple layer ejecta (MLE) based on the observed number of ejecta layers. (a) SLE craters display one complete ejecta layer around the crater, like this 9.9-km-diameter crater located at 19.54°S 277.01°E. (THEMIS image V18738005.) (b) DLE craters, like this 12.0-km-diameter example at 55.59°N 268.68°E, display two complete ejecta layers. (THEMIS image I13646007.) (c) MLE craters display three or more partial or complete ejecta layers. This MLE crater is 29.4 km in diameter and is located at 23.15°S 281.35°E. (THEMIS image I06994003.) (NASA/ASU.)

Table 5.2 *Diameter ranges of specific interior morphologies*

| Interior feature | Diameter range |
|---|---|
| Central peaks | ~6–175 km |
| Central pits | 5–60 km |
| Peak ring basins | ~50–500 km |
| Multi-ring basins | > 500 km |

*Source*: Data from Barlow (1988).

The excavation stage ends when the transient crater reaches its maximum size. For craters whose final shape and size are dictated by gravity rather than the strength of the target material, the time for transient crater formation is related to $D_t$ and $g$ (Melosh, 1989):

$$T \cong 0.54 \left( \frac{D}{g} \right)^{1/2}. \tag{5.12}$$

The modification stage extends from the end of transient crater formation until the crater is completely destroyed by subsequent geologic processes. During the modification stage, simple craters usually undergo some sliding of debris from the crater walls but most of the changes to simple craters result from infilling by other geologic processes such as volcanic, eolian, and fluvial activity. Complex craters display a wider variety of internal structures. As with ejecta structures, interior features tend to transition to different types as a function of crater size and location on the planet (Table 5.2). Martian complex craters often display a peak or pit in the center of the crater. Central peaks (Figure 5.12a) result from uplift of the floor after passage of the shock wave, with the uplift freezing into place to form the central peak. Larger craters can display a mountainous ring (peak ring) rather than a central peak complex (Melosh, 1989). The peak ring structure (Figure 5.12b) likely results from collapse of the central peak and formation of a "ripple" which freezes into place. The largest craters, such as Hellas and Argyre, may be multi-ring structures, although their outer rings are not obvious. Central pits are seen on Mars and on many of the icy moons in the outer Solar System. Martian central pits can occur either directly on the crater floor or on top of a central rise (Barlow, 2006) (Figure 5.12c). Central pits likely form from vaporization of ice under the central part of the crater floor, resulting from shock heating of the material and explosive release of the resulting gases (Wood *et al.*, 1978; Pierazzo *et al.*, 2005). The walls of complex craters typically have slopes greater than the angle of repose when the crater first forms, resulting in collapse of the walls to form terraces (Figure 5.12a).

Martian impact craters display a wide range of ejecta and interior morphologies, indicating complexities in target properties during crater formation and modification

(a)

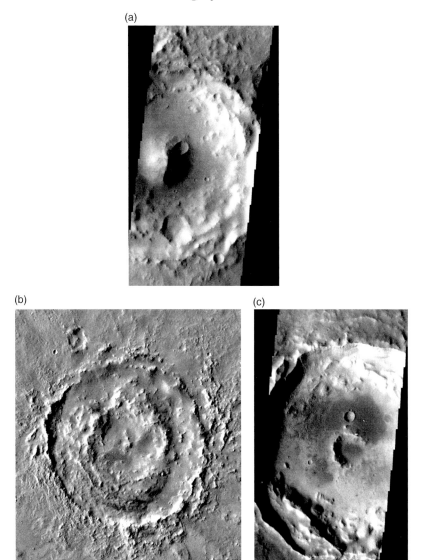

(b)

(c)

Figure 5.12 Interior morphologies vary with crater size and location. (a) Central peaks, such as the one visible in this 30.1-km-diameter crater at 8.30°N 302.54°E, are common features in complex impact craters. The collapse of the crater walls to form wall terraces can also be observed. (THEMIS image V18886012, NASA/ASU.) (b) Central peaks often expand into peak rings for larger craters. A peak ring is clearly visible in 224-km-diameter Lyot Crater, located at 50.38°N 29.33°E. (Viking mosaic image, NASA/JPL.) (c) Central pits are common in impact craters on Mars and on icy moons of the outer Solar System. This floor pit occurs in a 27.9-km-diameter crater located at 4.30°N 294.05°E. (THEMIS image V17526014, NASA/ASU.)

Figure 5.13 Pedestal craters are perched above the surrounding terrain by removal of fine-grained material by eolian and/or sublimation processes. These two pedestal craters are each about 1.5 km in diameter and are located near 52.0°N 150.5°E. (THEMIS image V11541006, NASA/ASU.)

processes operating after formation. For example, many small craters at high latitudes are elevated above the surrounding terrain (Figure 5.13). These "pedestal craters" formed when surrounding material was removed either by eolian deflation or ice sublimation (Barlow, 2006). Erosional processes have removed the elevated rims and infilled the floors of many older craters, particularly those found on the ancient Noachian surfaces. Computer simulations of the topography produced by different erosional processes have been compared to the topographic profiles across eroded craters. The results suggest that rainfall was a primary agent of erosion during the Noachian period (Craddock and Howard, 2002). Crater degradation in the Hesperian and Amazonian periods has been dominated by eolian and volcanic activity, with only localized fluvial/glacial erosion (Bibring *et al.*, 2006).

### 5.3.2 *Volcanism*

Volcanism can produce flat lava plains or a wide variety of topographic features called volcanoes. The type of volcanic feature produced during a volcanic eruption depends on the viscosity of the magma/lava involved. Viscosity depends on temperature, composition, presence/absence of solid material in the melt, and amount of gas dissolved in the magma. The most important factor influencing viscosity is the amount of silicate ($SiO_2$) in the magma – higher $SiO_2$ concentrations result in stickier magmas. High-viscosity magmas can contain more gas than lower-viscosity magmas and are therefore more explosive. Explosive eruptions also can occur when lower-viscosity magmas encounter water. Such eruptions are called phreatic eruptions and usually produce large circular craters called maars.

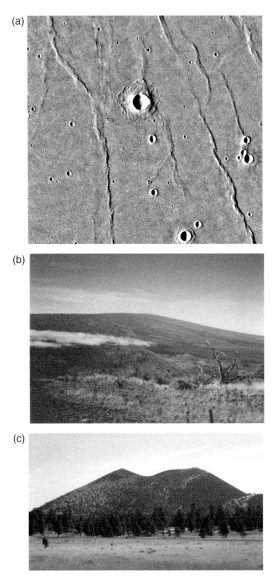

Figure 5.14 Magma viscosity and eruption rate determine the type of volcanic feature produced during an eruption. (a) Low-viscosity magmas often form flood basalts, as seen in this view of Lunae Planum, Mars, centered near 15.0°N 293.8°E. (Viking mosaic image, NASA/JPL.) (b) Slightly more viscous flows form low-relief constructs called shield volcanoes. Mauna Loa, Hawaii, is an example of a shield volcano. (Image by author.) (c) Cinder cones form when the gas content of the magma increases. The explosive release of this gas fragments the magma during the eruption, with these fragments cooling into cinders which form a conical structure. Sunset Crater in Arizona last erupted in AD 1064. (Image by author.) (d) Mt. St. Helens in Washington displays the classic steep-sloped profile of highly viscous composite volcanoes. (Image by author.) (e) This Space Shuttle radar image shows the Yellowstone caldera region. The heat associated with this region is revealed today through the area's active geysers and hot springs. Large eruptions occurred 2.0 Ma, 1.3 Ma, and 0.6 Ma ago, ejecting over $3700 \, \text{km}^3$ of material. (NASA/Shuttle/SIR-C/X-SAR.)

(e)

(d)

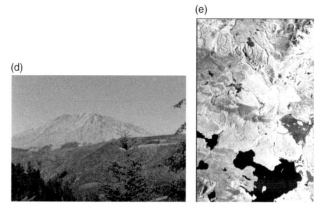

Figure 5.14 (cont.)

The type of volcanic feature produced during an eruption depends on viscosity and eruption rate. If two magmas have the same viscosity, the event with the higher eruption rate will generally produce a flatter structure than the event with the lower eruption rate.

Flat plains of lava flows are typically composed of basaltic rock (consisting of the minerals plagioclase, pyroxene, and olivine) and are called flood basalts (Figure 5.14a). Flood basalts are produced from low-viscosity magmas with high eruption rates. These eruptions have little to no associated explosive activity, resulting in very little topographic relief. The source of these massive eruptions is usually a vent or fracture. Flood basalts are a common form of volcanism, identified on all of the terrestrial planets as well as Earth's Moon.

Low-viscosity magmas produced by events with lower eruption rates result in topographically low structures called shield volcanoes (Figure 5.14b). These mountains have slopes typically <5° and are composed of basalt, although late-stage eruptions can be slightly more silicic due to differentiation within the volcano's magma chamber. Eruptions emanate from the central crater (caldera) or from fissures along the volcano's flanks. Shield volcanoes can be extremely massive – the volume of Mars' Olympus Mons shield volcano is about $3 \times 10^6 \, \text{km}^3$ (Smith *et al.*, 2001a), or about the same volume of basalt as is found in the entire Hawaiian–Emperor Seamount chain. Shield volcano eruptions produce basaltic lava flows, which can be rough (aa) or smooth (pahoehoe) (Figure 5.15). Maps of surface roughness derived from MOLA analysis suggest that both pahoehoe and aa flows occur on Mars.

Fluid lavas producing flood basalts and shield volcanoes often move from one location to another through either lava channels on the surface or underground in lava tubes (Figure 5.16). Once the lava reaches a flat plains region, it will spread out along the sides into a large flow. The amount of spreading depends on the properties

(b)

(a)

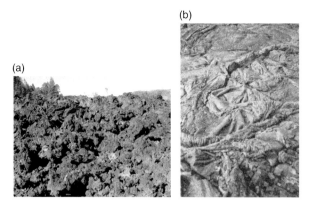

Figure 5.15 Low-viscosity lavas produce two types of lava flows. (a) Rougher flows are called aa flows, as seen in this image of the Bonito Lava Flow at Sunset Crater, Arizona. (b) Smoother flows, like these on Kilauea Volcano in Hawaii, are called pahoehoe flows. Pahoehoe flows are characterized by a ropy texture. (Images by author.)

Figure 5.16 Low-viscosity lavas often form lava channels and lava tubes. These lava channels occur on the flank of Pavonis Mons. This is a perspective view of the region taken by MEx's HRSC camera. (Image SEMELC9ATME, ESA/DLR/FU Berlin [G. Neukum].)

of the lava, especially the viscosity and the amount of solid material contained in the flow. Lava flows are characterized by their aspect ratio ($A$):

$$A = \frac{t}{w} \tag{5.13}$$

where $t$ is the central thickness of the flow and $w$ is the flow's width. The value of $A$ depends on slope, $g$, and viscosity. As slope decreases, $A$ decreases since the lava flow spreads out, decreasing $t$ and increasing $w$. Conversely, $A$ increases as viscosity increases since the increased stickiness of the lava will reduce spreading. Comparing the aspect ratios of different lava flows provides important information about topographic variations and the properties of the originating magmas.

Eruptions become more explosive as the silica content increases. The release of gas trapped in the magma by the $SiO_2$ causes lava fountains. Lava fountain activity erupts small clumps of magma into the air, which then cool into cinders as they fall back to the surface. The accumulation of cinders creates a cone whose flanks have slopes close to the angle of repose (about 30° for cinders). These cinder cones (Figure 5.14c) are primarily basaltic in composition with cinder formation from a central caldera and lava flows often exuded from their base. Cinder cones commonly form on the flanks and later in dormant calderas of shield volcanoes as the magma chamber composition evolves.

Composite or stratovolcanoes are steep-sided volcanic constructs composed of alternating layers of ash and lava flows (Figure 5.14d). They occur in association with subduction zones on Earth, but have not been definitively identified on other planets. Stratovolcanoes contain very viscous and volatile-rich magmas, resulting in extremely explosive eruptions from the central caldera. Sticky lava domes often form in the caldera, plugging the conduit used by the magma to reach the surface. Pressure builds up under this dome until the gases can break through the surface, producing a sudden release of gas and ash. Hot clouds of ash can travel down the volcano flanks at speeds of hundreds of kilometers per second. These pyroclastic flows are the cause of much death and destruction associated with stratovolcano eruptions. However, stratovolcanoes can also experience quieter extrusions of lava flows, giving rise to the stratified structure. After an eruption, a lava dome begins to rebuild in the caldera, allowing the process to repeat itself.

The most explosive eruptions are ash flows called ignimbrites. The flow consists of hot gas and ash particles which cover large areas. These eruptions result from extremely viscous magmas with acidic compositions. On Earth they only occur in continental regions, usually leaving behind a large caldera which can be mistaken for a simple valley (Figure 5.14e).

Mars displays a range of volcanic features ranging in age from Noachian to late Amazonian, with recent volcanic activity concentrated in the Tharsis and Elysium regions (Hodges and Moore, 1994). Flood basalts are common as lava plains surrounding the big volcanoes in Tharsis and Elysium and in the ridged plains. Large expanses of intercrater plains seen between craters in the Noachian-aged southern highlands are probably ancient flood basalts. Flood basalts appear to have occurred throughout martian history (McEwen *et al.*, 1999), although their areal extent has decreased over time in conjunction with the overall decline in volcanic activity.

Mars displays a number of very large but extremely low-relief (average slopes <3°) volcanoes called paterae (Latin for dish or saucer) (Figure 5.17a). Paterae are the

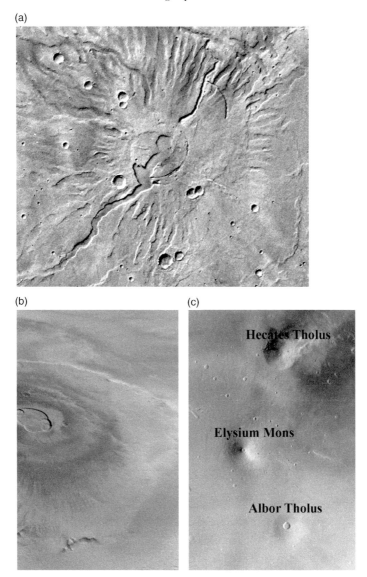

Figure 5.17 Various types of volcanic structures occur on Mars. (a) Tyrrhena Patera is a very low relief, highly eroded, and heavily cratered volcano centered near 21.5°S 106.5°E. (Viking mosaic image, NASA/JP.) (b) Olympus Mons is a huge shield volcano, 21.3 km high and ~800 km in diameter. (NASA/MSSS.) (c) The Elysium Mons province contains three volcanoes, Hecates Tholus, Elysium Mons, and Albor Tholus. (NASA/MSSS.)

oldest volcanic structures on Mars and are primarily concentrated around the Hellas impact basin. Detailed analysis of Hadriaca, Tyrrhena, and Apollinaris Paterae indicate the main edifices formed by explosive volcanism during the Late Noachian and Early Hesperian periods, with subsequent effusive eruptions extending until the Early Amazonian (Crown and Greeley, 1993; Robinson *et al*., 1993; Berman *et al*., 2005). Paterae flanks are highly dissected by wind and/or fluvial activity. The ease with which these flanks have been eroded suggests they are composed of fine-grained pyroclastic deposits, either in the form of pyroclastic flows or air-fall ash deposits. However, paterae compositions are consistent with low-viscosity basalt. Mafic (iron-rich) pyroclastic deposits can be produced either by water–magma interactions or by rapid ascent of deep-sourced magmas, and both mechanisms have been proposed for the martian paterae (Crown and Greeley, 1993; Gregg and Williams, 1996).

The low-albedo region of Syrtis Major is a low-relief (slopes $<1°$) volcano of basaltic composition (Bandfield *et al*., 2000; Hoefen *et al*., 2003) rising up to an elevation of $\sim$2.3 km (Hiesinger and Head, 2004). Schaber (1982) argued that Syrtis Major is a shield volcano with two calderas (Nili and Meroe Paterae), but Hodges and Moore (1994) and Hiesinger and Head (2004) argue that its low relief makes it more akin to the highland paterae. The Early Hesperian age of Syrtis Major places it in the same time period during which the highland paterae were forming, suggesting that during this period eruptions were too temporally limited, lava viscosities were too low, or the rate of eruption was too high to build larger edifices (Hodges and Moore, 1994). The volume of lava comprising Syrtis Major is estimated to be $\sim$1.6 to $3.2 \times 10^5$ km, comparable to estimated volumes of Amphitrites Patera (Hiesinger and Head, 2004).

Mars' big shield volcanoes were first revealed by Mariner 9 as circular depressions atop high mountains which poked through the global dust storm. As the dust settled, these features were recognized as the calderas topping the four massive shield volcanoes in the Tharsis region: Olympus Mons (Figure 5.17b), Ascraeus Mons, Pavonis Mons, and Arsia Mons. Olympus Mons rises to a height of 21.3 km, while the elevations of the other three shields range between 14.1 and 18.2 km (Smith *et al*., 2001a). Elysium Mons, the largest shield volcano in the Elysium province (Figure 5.17c), rises to an elevation of 14.1 km.

Smaller volcanic edifices with slightly higher slopes (up to 12°) are called domes or tholii (singular is tholus). These features appear to be small shield volcanoes based on their large calderas and other geomorphic features (Plescia, 1994, 2000).

The large and small shields are concentrated in the Elysium and Tharsis regions of Mars. The Elysium province contains three volcanoes: Albor Tholus, Hecates Tholus, and Elysium Mons (Figure 5.17c). Albor Tholus, 160 km in diameter, is a small shield probably composed of basalt. Crater analysis suggests a Noachian to

Hesperian age for this volcano (Hodges and Moore, 1994). Hecates Tholus is about 200 km in diameter, has a relatively small caldera, and displays a large number of channels on its flanks (Gulick and Baker, 1990; Mouginis-Mark and Christensen, 2005). The volcano is primarily Hesperian in age (Hodges and Moore, 1994), although HRSC crater data have been interpreted to indicate Amazonian activity on the northwestern flank (Neukum *et al.*, 2004). The western flank of Hecates Tholus displays lower crater density and has been proposed to be a younger pyroclastic deposit from either a summit (Mouginis-Mark *et al.*, 1982) or flank (Hauber *et al.*, 2005) eruption.

Elysium Mons is the largest of the Elysium volcanoes, with a diameter of 400 km and a height of 14 km. It has a single summit caldera from which emanate short lava flows (<70 km long). Lava flows originating from flank vents are much longer, up to 250 km long, and display large variations in aspect ratio (Mouginis-Mark and Yoshioka, 1998). The flows likely originate as low-viscosity lavas, but viscosity increases exponentially with distance as degassing occurs and temperature drops (Glaze *et al.*, 2003). Crater analysis suggests that Elysium Mons is the youngest of the Elysium Province volcanoes with a Late Hesperian to Early Amazonian age (Hodges and Moore, 1994).

The Tharsis region has the largest concentration of structures on Mars with 12 large volcanoes, many smaller features, and extensive lava flows (Figure 5.18). The Tharsis Bulge occupies about 25% of the surface and rises about 6 km above the mean planetary radius (Anderson *et al.*, 2001; Phillips *et al.*, 2001; Smith *et al.*, 2001a). Gravity data suggest that this bulge has been largely constructed from voluminous lava flows, although some support by a mantle plume is also suggested (Smith *et al.*, 2001a; Phillips *et al.*, 2001; Kiefer, 2003). Crater and tectonic data indicate that much of martian volcanism was localized in the Tharsis region as early as the end of the Middle Noachian period (Solomon *et al.*, 2005).

The Tharsis Bulge is capped by the three shield volcanoes Ascraeus, Pavonis, and Arsia Montes. Ascraeus, Pavonis, and Arsia are aligned along a northeast-to-southwest rift zone about 700 km apart. Long, narrow lava flows, indicative of high eruption rates, extend from these volcanoes. The three volcanoes are believed to have originated during the Noachian period with slow accumulation of fluid lavas from the summit calderas and surrounding concentric fissures. Multiple calderas often seen on the martian shields suggest several episodes of filling and withdrawal from the magma chamber. After the shields reached their maximum heights, the eruptions shifted to the northeast–southwest rift zone, producing younger flank eruptions. The low aspect ratios of the flow and presence of lava channels indicate very fluid lavas. Smaller shield volcanoes are seen on the flanks and caldera of Ascraeus Mons, suggesting the presence of large dike complexes within the larger shields (Wilson and Head, 2002). Flood basalts formed throughout the Tharsis

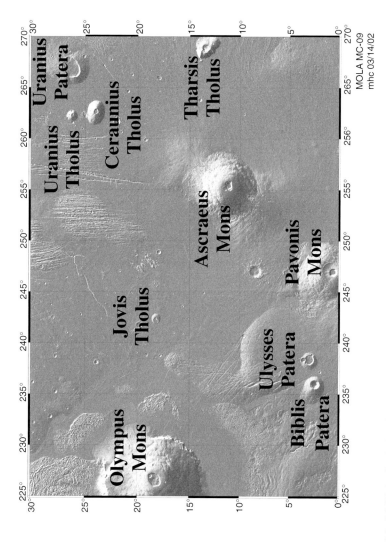

Figure 5.18 This MOLA shaded relief map shows most of the volcanoes associated with the Tharsis region. (Image courtesy of Michael H. Carr and MOLA Science Team.)

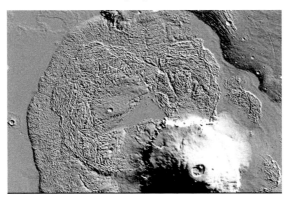

Figure 5.19 The Olympus Mons aureole consists of blocks, troughs, and ridges and extends to the northwest of the Olympus Mons volcano. Both volcanic and tectonic origins have been proposed for the aureole. (MOLA shaded relief image, NASA/GSFC.)

region, originating from vents and fissures recently identified in MOC and THEMIS imagery (Mouginis-Mark and Christensen, 2005). Ages based on crater counts suggest long-lived activity among the Tharsis shield volcanoes, with some lava flows perhaps as recent as 40 Ma ago (Hartmann *et al.*, 1999; Neukum *et al.*, 2004).

MOLA analysis reveals that Olympus Mons lies on the western edge of the Tharsis uplift, so its evolution is largely separate from that of the Tharsis Bulge. Both summit and vent eruptions contributed to the formation of this huge edifice (Mouginis-Mark and Christensen, 2005). The shield has several calderas at its summit, indicating multiple episodes of eruption and collapse. Ages of the calderas are estimated to be between 100 and 200 Ma (Neukum *et al.*, 2004). Olympus Mons is 600 km in diameter and is partially surrounded by a scarp up to 10 km in extent. The Olympus Mons aureole, a region of blocks, arcuate ridges, and deep extensional troughs, lies about 300–700 km northwest of the scarp (Figure 5.19). Possible origins for the aureole include failure of the volcano's flanks and volcanic products ranging from pyroclastic flows to eroded lava flows. Analysis of MGS and MO data supports the idea that flank failure and mass movement produced the basal scarp and aureole deposits (McGovern *et al.*, 2004b).

The other large volcano in the Tharsis region is 1600-km-diameter Alba Patera, located on the northern edge of the Tharsis uplift (Figure 5.20). Alba Patera is surrounded by a circumferential system of extensional faults (graben). The summit caldera rises to 6.8 km height and is offset to the southwest of the center of this faulting. MOLA-derived slopes are <2°. Although most models suggest that Alba Patera is composed of low-viscosity basaltic lava flows (Schneeberger and Pieri, 1991), possible pyroclastic deposits have also been identified on the volcano's flanks (Mouginis-Mark *et al.*, 1988). Many models propose that the circumferential

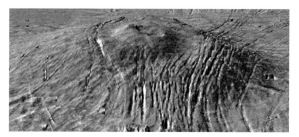

Figure 5.20 Alba Patera is 1600 km in diameter and is surrounded by circumferential graben. This three-dimensional view of the volcano is obtained by draping Viking imagery over MOLA topography. (NASA/GSFC/MOLA Science Team.)

graben resulted from lithospheric stresses produced by the volcano's surface load (Comer *et al.*, 1985; Turtle and Melosh, 1997). However, McGovern *et al.* (2001) argue that sill complexes within the lithosphere are more consistent with the MGS-derived topography and gravity of this region.

The remaining Tharsis volcanoes consist of small shields largely built by effusive basaltic eruptions. Three lie on the western flank of the Tharsis Bulge (Jovis Tholus, Ulysses Patera, and Biblis Patera) while the other four are grouped on the northeastern flank (Ceraunius, Uranius, and Tharsis Tholii and Uranius Patera) (Figure 5.18). The smaller sizes of these shields indicate lower volumes of erupted material compared to the larger shields. The volcanoes are composed primarily of basaltic flows, although late-stage pyroclastic activity has been suggested for Ceraunius and Uranius Tholii (Plescia, 2000). Crater analysis suggests that all seven small shields were formed within $10^4$–$10^5$ years during the Hesperian (Plescia, 1994, 2000).

A thick deposit of fine-grained, easily eroded material is found southwest of the Tharsis shields (Figure 5.21). The origin of this material, called the Medusae Fossae Formation (MFF), is controversial, with dust deposits, ancient polar layered deposits, and pumice rafted on an ancient ocean among the proposed origins. The most widely accepted origin for these deposits is that they are ash deposits (Scott and Tanaka, 1982; Edgett *et al.*, 1997). The lower atmospheric pressure on Mars may allow basaltic eruptions to be more explosive than is typically observed on Earth (Wilson and Head, 1994), giving rise to basaltic ash deposits. The detection of high concentrations of chlorine coincident with the MFF also is consistent with a volcanic origin (Taylor *et al.*, 2006). Layered deposits in Valles Marineris, Terra Meridiani, and Arabia Terra also have been suggested to be pyroclastic deposits from explosive volcanic eruptions in Tharsis (Hynek *et al.*, 2003).

The youngest volcanic unit on Mars lies between the Tharsis and Elysium provinces in the southwestern part of Amazonis Planitia, centered near 30°N 200°E

Figure 5.21 The Medusae Fossae Formation southwest of the Tharsis region is composed of fine-grained materials, as indicated by the range of wind-eroded features. It may be an ash deposit. (THEMIS image I00968002, NASA/ASU.)

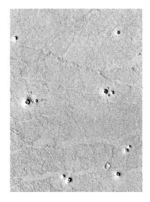

Figure 5.22 Small cones are seen in several locations across Mars. These are found in Amazonis Planitia near 24.8°N 188.7°E. The image is about 3 km across and the diameters of the small cones are < 250 m. (NASA/JPL/MSSS/University of Arizona.)

with an areal extent of $\sim 10^7\,\mathrm{km}^2$ (Plescia, 1990; Keszthelyi *et al.*, 2000). Ages derived from crater analysis suggest that flood volcanism occurred in this region as recently as 10–100 Ma ago (Hartmann and Berman, 2000; Berman and Hartmann, 2002), although HRSC analysis suggests volcanism as recently as 3 Ma ago (Werner *et al.*, 2003). The source region(s) of these flows have been identified as the Cerberus Fossae fracture system and/or small shields on the western side of Cerberus Planitia (Plescia, 1990, 2003; Keszthelyi *et al.*, 2000; Werner *et al.*, 2003).

Many small cones (4–16-km-diameter) are seen on Mars (Figure 5.22). Hodges and Moore (1994) proposed that these features were formed by water–magma interactions, although MOLA analysis by Garvin *et al.* (2000b) led them to suggest that these features are small shields built by effusive lava flows. Other small cones around the planet have been attributed to phreatomagmatic interactions (e.g., Lanagan *et al.*, 2001).

### 5.3.3 Tectonism

Tectonic activity includes any crustal deformation caused by surface motion. Planetary tectonism is a result of stress and strain within the rigid lithosphere. Stress ($\sigma$) is the force per area exerted on material while strain ($\varepsilon$) is the measure of how much the material deforms when the stress is applied. An elastic material deforms when a force is applied to it, but returns to its original shape when the force is removed. Elastic materials deform according to Hooke's law:

$$\sigma = \varepsilon_e E. \tag{5.14}$$

Hooke's law shows that the ratio of the stress $\sigma$ and the elastic strain $\varepsilon_e$ is equal to a property of the material called Young's modulus ($E$) which measures the "stiffness" of the material.

Alternately, viscous material remains deformed when the applied force is removed. In a simple Newtonian fluid, the rate of change of fluid strain ($\varepsilon_f$) is directly proportional to the applied stress:

$$\dot{\varepsilon}_f = \frac{d\varepsilon_f}{dt} = \frac{\sigma}{2\eta}. \tag{5.15}$$

The response of a viscous material to an applied stress depends on the viscosity ($\eta$) of the material. Most geologically important materials are neither purely elastic nor viscous, but a combination. Viscoelastic materials do not completely return to their pre-deformation appearance but they also do not stay completely deformed when the stress is removed. The viscoelastic strain ($\varepsilon_v$) is the sum of the viscous and elastic terms:

$$\varepsilon_v = \varepsilon_e + \varepsilon_f. \tag{5.16}$$

Therefore

$$\dot{\varepsilon}_v = \left(\frac{1}{E}\right)\frac{d\sigma}{dt} + \left(\frac{1}{2\eta}\right)\sigma. \tag{5.17}$$

Viscoelastic materials can approach their initial shape over some period of time called the viscoelastic relaxation time ($\tau_v$). By setting $\dot{\varepsilon}_v = 0$ and integrating Eq. (5.17), we find that

$$\sigma = \sigma_0 e^{\frac{-Et}{2\eta}} \tag{5.18}$$

Figure 5.23 The presence of near-surface ice weakens the crustal material, allowing viscoelastic relaxation to occur. Features in the mid-latitude regions of Mars often display rounded edges, like the rims of these two craters. This terrain softening is indicative of near-surface ice which can become warm enough to cause such topography to deform. The image is centered near 43.7°S 357.4°E and the larger crater is approximately 36 km in diameter. (THEMIS image I07166004, NASA/ASU.)

where $\sigma_0$ is the initial stress and $t$ is the time. The stress relaxes to a value of $1/e$ times its original value in $\tau_v$:

$$\tau_v = \frac{2\eta}{E}. \qquad (5.19)$$

Viscoelastic behavior of the ice-rich martian crust leads to relaxation of mid-latitude topography, including craters and mesas, in a process called terrain softening (Figure 5.23) (Squyres and Carr, 1986).

A viscoelastic material responds elastically to a stress applied on a timescale which is short relative to the viscoelastic relaxation time and viscously to a stress applied on a timescale which is long compared to $\tau_v$. The material's temperature is an important consideration when determining how it will respond to an applied stress since the material can be brittle at low temperatures and ductile at high temperatures. Planetary lithospheres behave elastically in response to typical stresses. If the applied force is too strong, the material will fracture, creating a fault. The deeper layers of the asthenosphere behave viscously to applied forces and will exhibit ductile behavior.

The strength of the applied force determines whether lithospheric rocks will fold (bend) or fault (fracture). Materials will experience ductile deformation up to some level of $\sigma$ and $\dot{\varepsilon}$ during which folding will occur. Once the stress and/or strain rate become too large, the material experiences brittle deformation and faulting results. If the crust is cracked because it moved in response to stresses in the crust, the cracks

Figure 5.24 Extensional stresses cause downdropped valleys called graben. These graben occur south of the Alba Patera volcano. The crater at upper right is 2.7 km in diameter and located at 36.1°N 255.8°E. (THEMIS image I11999006, NASA/ASU.)

are called faults. However, if the crust is cracked due to stresses but no crustal motion occurred, the cracks are called joints.

In 1905 E. M. Anderson realized that tectonic features on Earth resulted from variations in stress orientations. Faults are the mechanism by which the crust is extended in one direction and compressed in the direction 90° from the extension. Anderson's theory of faulting notes that there are three principal stress directions and that the type of tectonic features which results depend on the relative orientations of those principal stress directions. Two of the principal stress directions lie horizontally within the lithosphere while the third is perpendicular to the planet's surface. The three principal stress directions correspond to the maximum ($\sigma_1$), intermediate ($\sigma_2$), and minimum ($\sigma_3$) stresses.

Extensional stresses will pull the surface apart and produce normal faults, where material on one side of the fault will be downdropped relative to the other side. Normal faults occur when $\sigma_2$ and $\sigma_3$ lie within the lithosphere and $\sigma_1$ is vertical. Many valleys in the Basin and Range Province of the southwestern United States are the result of downdropped blocks (graben) bordered by normal faults. Grabens are common extensional features seen around the Tharsis Bulge on Mars (Figure 5.24).

Compressional stresses push portions of the crust together, creating thrust faults where material on one side of the fault is uplifted relative to the other side. Thrust faults form when $\sigma_1$ and $\sigma_2$ lie within the lithosphere and $\sigma_3$ is vertical. Blocks bounded by thrust faults which have been uplifted relative to their surroundings are called horsts. Wrinkle ridges, common on the martian ridged plains (Figure 5.14a), are compressional features which typically result from sagging of the crust under the weight of large expanses of lava flows.

Strike-slip faults result when $\sigma_1$ and $\sigma_3$ lie within the lithosphere and $\sigma_2$ is vertical. This results in the crustal blocks sliding past one another. If you stand on one side of the strike-slip fault and look across to the opposite crustal block, you classify the fault as a right-lateral fault if the opposite crustal block moves to your right. The fault is a left-lateral fault if the opposite crustal block moves to the left.

Planets with plate tectonics clearly display these three fault classes at plate boundaries. Divergent boundaries, where two plates diverge, display extensional faults while convergent boundaries, where plates converge, display compressional faults. Strike-slip faults occur at transform boundaries where plates slide past each other. Plate tectonics on Mars is still a controversial topic (Section 2.4.2), but such activity would have been limited to very early martian history if it occurred. While extensional, compressional, and strike-slip faults are observed on Mars, none of the geologic features definitively indicate current plate tectonic activity. Localized stresses and strains, rather than global plate tectonics, have dominated the tectonic history of Mars.

The level of current tectonic activity on Mars is unknown because of the lack of seismic data (Section 3.3). Indications of recent volcanism in the Amazonis Planitia and Tharsis regions suggest that seismic activity should occur in those locations. Transection relationships reveal that extensional and compressional tectonism has occurred throughout martian history, with activity primarily concentrated around Elysium and Tharsis (Anderson *et al.*, 2001, 2006). Anderson *et al.* (2001) determined that tectonic activity in the Tharsis region has occurred throughout martian history with the centers of the activity shifting over time (Figure 5.25). Tectonic activity associated with Elysium has occurred only recently. Noachian-aged faulting was centered in the Claritas region near 27°S 254°E while Late Noachian–Early Hesperian faulting was concentrated along the margins of Syria, Sinai, and Solis Plana. Early Hesperian graben and wrinkle ridges are centered on Syria Planum and Tempe Terra. Extensional faults radial to Alba Patera dominate in the Late Hesperian–Early Amazonian. The most recent tectonic activity in the Tharsis region has centered on the large volcanic shields, with a center near the southern flank of Ascraeus Mons (8°N 200°E). Elysium activity is concentrated in the Middle to Late Amazonian.

The largest extensional feature on Mars is the Valles Marineris canyon system (Figure 5.26), which stretches along the equator for ~4000 km between 250°E and 330°E (Lucchitta *et al.*, 1992). Parts of the canyon are up to 6 km below the 0-km elevation contour and up to 11 km below the surrounding plains (Smith *et al.*, 2001a). The canyon is divided into three segments based on morphologic changes. The western end of the canyon consists of a series of interconnected canyons called Noctis Labyrinthus. The central portion consists of roughly east–west trending canyons extending for ~2400 km. The eastern section contains irregular depressions which merge with the chaotic terrain and outflow channels.

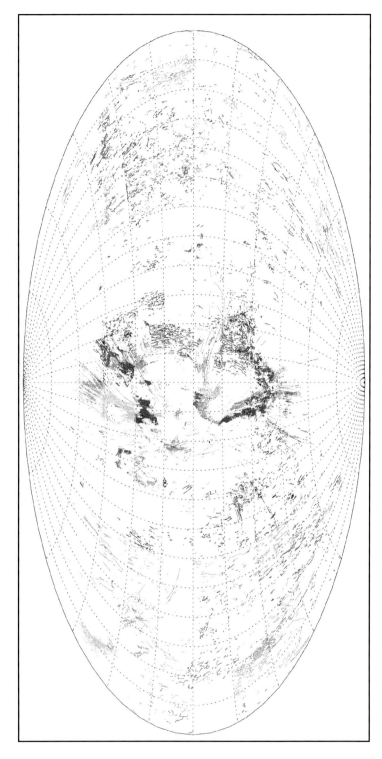

Figure 5.25 This map shows all the tectonic structures visible on Mars. Map is centered at 0°N 270°E, near the Tharsis tectonic center. (Image courtesy of Robert Anderson, JPL.)

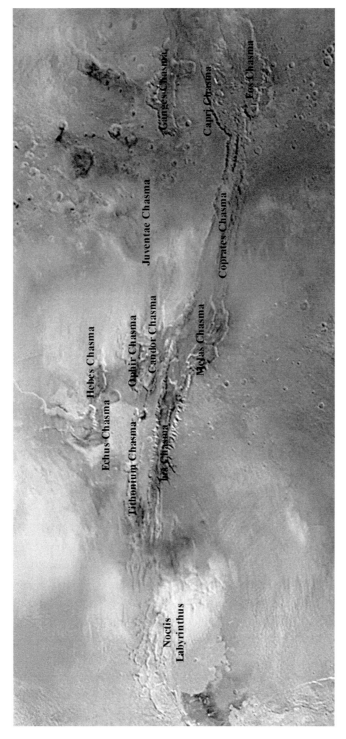

Figure 5.26 Valles Marineris is composed of a series of smaller canyons. The Valles Marineris system extends over 4000 km near the equator. (MOC image MOC2-144, NASA/JPL/MSSS.)

(a)                                        (b)

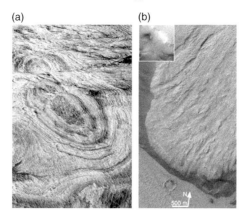

Figure 5.27 The interior of Valles Marineris shows layered rocks and landslides, providing insights into its geologic evolution. (a) Layers of bright and dark rocks are seen throughout the canyon. This exposure occurs in the western part of Candor Chasma. The image is ~3 km across and located at 5.7°S 284.2°E. (MOC image MOC2-682, NASA/JPL/MSSS.) (b) Landslides are common within the Valles Marineris canyons. This image shows the toe of a landslide in Ganges Chasma, near 8.0°S 315.6°E. (MOC image MOC2-295, NASA/JPL/MSSS.)

Valles Marineris originated during the Late Noachian or Early Hesperian either from dike emplacement associated with Syria Planum (Mège and Masson, 1996) or stresses associated with the uplift of Tharsis (Smith *et al.*, 2001a). Subsequent subsidence and normal faulting have continued into the Amazonian (Schultz, 1998). MOLA analysis reveals that the canyon is deepest in Coprates Chasma near 300°E, with the western and eastern sections dipping towards that location. The eastern slope of ~0.03° has existed since canyon formation. Water could only have flowed eastward out of the main canyon if the water depth was >1 km to overcome the observed topography (Smith *et al.*, 1999).

The canyon exposes stratigraphic layers emplaced throughout martian history (Figure 5.27a). Thin (tens of meters) strong layers are interspersed with thicker (hundreds of meters) weak layers (Beyer and McEwen, 2005). The strong layers are probably lava flows (McEwen *et al.*, 1999; Williams *et al.*, 2003), but the weaker layers could be sedimentary (Malin and Edgett, 2000a), aqueously altered products (Treiman *et al.*, 1995), or thin/weak volcanic products (Beyer and McEwen, 2005).

The layers are the source of the numerous landslides displayed in the canyon (Figure 5.27b). These landslides are examples of mass wasting processes caused by the martian gravity exerting a downward force on precariously supported materials. Quantin *et al.* (2004) use crater analysis to determine that the landslides have been occurring since 3.5 Ga ago, with the youngest slides displaying ages of ≤50 Ma. Both dry granular flows (Soukhovitskaya and Manga, 2006) and wet flows (Harrison and Grimm, 2003) match the morphologies of Valles Marineris landslides,

Figure 5.28 Giant polygons are seen in Utopia and Acidalia Planitiae. They are formed by extensional stresses resulting from uplift or compaction of overlying sediments. These polygons are located in Utopia Planitia near 47.0°N 129.2°E. Image is ~30 km in width. (THEMIS image I10119010, NASA/ASU.)

suggesting both mechanisms have operated in different time periods and locations throughout the canyon.

Extensional stresses also are implicated in the formation of giant polygons observed in the northern plains, particularly in Acidalia and Utopia Planitiae (Figure 5.28). These features range in diameter from 2 to 32 km and are bounded by graben with widths between 0.5 and 7.5 km and depths between 5 and 115 m (Hiesinger and Head, 2000). Although a variety of formation models have been proposed for these features, the two dominate models are the tectonic and drape-fold models. The tectonic model proposes that giant polygons result from uplift of basins following removal of large bodies of liquid water (Hiesinger and Head, 2000; Thomson and Head, 2001). The drape-fold model argues that sedimentary layers (of possible lacustrine origin) deposited over rough topography will lead to differential compaction of sediments and produce polygonal features on scales comparable to the underlying topographic variations (McGill and Hills, 1992; Buczkowski and McGill, 2002).

MOLA analysis of the eastern hemisphere dichotomy boundary reveals both normal and thrust faults, indicating that formation of the dichotomy boundary involved both extensional and compressional stresses (Watters, 2003). Compressional stresses occurred in many areas of Mars, as indicated by the widespread distribution of wrinkle ridges. Wrinkle ridges are linear broad arches with a superposed ridge (Figure 5.29). They are typically a few tens of kilometers in width and are 80 to 300 m high (Golombek *et al.*, 2001). Wrinkle ridges are now generally accepted to be surface expressions of subsurface thrust faults (Schultz, 2000; Golombek *et al.*, 2001; Watters, 2004; Goudy *et al.*, 2005), although buried impact craters sometimes contribute to wrinkle ridge patterns. Wrinkle ridges are most

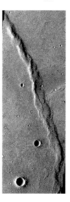

Figure 5.29 Wrinkle ridges form from compressional stresses and are often seen on large expanses of flood basalts. This wrinkle ridge on Lunae Planum, near 13.8°N 296.6°E, exhibits the typical morphology of a broad arch with a superposed ridge. The crater near the bottom is ~1.9 km in diameter. (THEMIS image V14119007, NASA/ASU.)

frequent on the Hesperian-aged flood basalts called ridged plains. Buried Hesperian-aged volcanic flows, such as that under the northern Vastitas Borealis formation, also show evidence of wrinkle ridges in MOLA analysis (J. W. Head *et al.*, 2002), indicating large-scale compressional folding and faulting of these units.

### 5.3.4 Mass movement

Mass movement features are produced when gravity causes collapse. The rate of collapse can be rapid (as with avalanches) or slow (as in subsurface creep). Solid rock is more stable than unconsolidated materials, but removal of underlying support (such as caused by erosion near the base of a cliff) can result in bedrock undergoing collapse. Addition of water or ice to soils reduces the strength of the material and enhances mass movement, such as happens when heavy rains cause mudslides on Earth.

Pouring sand onto a surface results in the sand forming a conical pile. The angle between the surface and the cone's sides is the angle of repose and is related to the slope of the cone's sides. Fine-grained material has a lower angle of repose (typically around 30°) than coarser material. Angular fragments can form cones with steep slopes.

Whenever the slope exceeds the angle of repose for a block of material, that block will move downslope under the influence of gravity. Rocks along a cliff can fall, producing a rock-fall. Large blocks of bedrock can slide along cracks and joints to create rockslides. Unstable soil will produce landslides. The collapse of crater walls to produce wall terraces (Figure 5.12a) and the landslides seen within

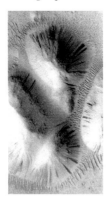

Figure 5.30 Dark slope streaks are caused by dust avalanches occurring on steep slopes. The bright dust slides downslope, revealing the darker underlying material. These dark streaks occur on buttes in the Aeolis region near 1.5°S 157.1°E. Image is ~3 km across. (MOC image MOC2-1439, NASA/JPL/MSSS.)

the Valles Marineris canyons (Figure 5.27b) are examples of rapid mass movement on Mars. Dark streaks (Figure 5.30) seen on features with slopes exceeding the angle of repose are produced by avalanches of unconsolidated material (Sullivan *et al.*, 2001).

Mass movement rates can also be very slow, as exhibited in soil creep. Soil creep is downhill movement of soil that occurs due to freeze–thaw, wet–dry, and thermal expansion cycles. It is imperceptible over short time periods, but can affect topography over long time periods. The rounding of sharp topography in the mid-latitudes of Mars is an example of soil creep caused by the presence of ice (Figure 5.23) (Squyres and Carr, 1986). Debris aprons at the base of bedrock escarpments also are proposed to result from creep of ice-rich materials (Perron *et al.*, 2003).

### 5.3.5 Eolian features

Planets with atmospheres show the effects of eolian or wind processes. Wind transports material from one location to another and causes both deposition and erosion. The physics of fluid dynamics is applied to material being transported by the wind (Greeley and Iverson, 1985). Larger material is transported by traction, the rolling of material along the surface. Slightly smaller material can bounce along the surface, a process called saltation. Pebbles are sometimes moved by impact creep, where saltating grains impart momentum through their impact on the pebble. The smallest material is carried within the wind flow by suspension. Depending on the wind speed and thickness of a planet's atmosphere, suspension typically operates on particles $\leq 60\,\mu m$ in diameter (dust), saltation efficiently moves material in the

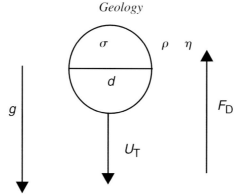

Figure 5.31 A particle of diameter *d* and density $\sigma$ falling through an atmosphere of density $\rho$ and viscosity $\eta$ experiences a downward force due to gravity *g* and an upward drag force $F_D$.

60–2000 µm size range (sand), and creep and traction move particles >2000 µm in diameter (granules and pebbles).

A particle of diameter *d* and density $\sigma$ falling through quiescent air of density $\rho$ and viscosity $\eta$ will feel a downward force due to gravity (*g*) and an upward force due to aerodynamic drag (Figure 5.31). The downward gravitational force is

$$F_g = mg = \left(\frac{4}{3}\right)\pi\left(\frac{d}{2}\right)^3 \sigma g = \left(\frac{1}{6}\right)\pi d^3 \sigma g. \qquad (5.20)$$

The upward drag force is obtained by using Stokes' drag equation for a spherical particle moving through a fluid:

$$F_D = 3\pi\eta U_T d. \qquad (5.21)$$

A particle dropped from rest will accelerate downwards under the influence of gravity until the drag force equals its weight. After that it moves at a constant velocity called the threshold velocity ($U_T$). For small particles, the drag is proportional to $U_T$, while for large particles or in atmospheres with low $\eta$ the drag increases as $U_T^2$.

Atmospheres, like fluids, can be gently flowing (laminar regime) or very turbulent. Reynolds' number ($R_e$) dictates whether the particle is moving in the laminar or turbulent regime:

$$R_e = \frac{\rho U_T d}{\eta}. \qquad (5.22)$$

Laminar regimes have low $R_e$, but turbulence dominates when $R_e \geq 1000$. Laminar flow typically occurs close to the planet's surface. Experiments find that $R_e \approx 24/C_D$

in the laminar regime, where $C_D$ is the drag coefficient of the gas or fluid. Substituting this value for $R_e$ in Eq. (5.22) gives an expression of the viscosity of the atmosphere:

$$\eta = \frac{C_D \rho U_T d}{24}.$$ (5.23)

This can be substituted back into Eq. (5.21):

$$F_D = \frac{C_D \pi \rho U_T^2 d^2}{8}.$$ (5.24)

Equating the gravitational force (Eq. 5.20) and the drag force (Eq. 5.24) allows us to determine the threshold velocity:

$$U_T = \left(\frac{4\sigma g d}{3\rho C_D}\right)^{1/2}.$$ (5.25)

The force ($F_s$) that the atmosphere must apply to a particle to keep it suspended is the weight of the particle reduced by the difference in particle and atmospheric density:

$$F_s = \left(\frac{1}{6}\right)\pi d^3 (\sigma - \rho)g.$$ (5.26)

To determine how a particle will be transported by the wind, $U_T$ is compared to the friction speed ($v_*$). The friction speed is not a true wind speed but is approximately equal to the vertical component of the particle velocity near the surface when the particle is experiencing atmospheric turbulence. The friction speed is related to the shear stress experienced by the atmospheric flow near the surface ($\tau$) and the atmospheric density ($\rho$):

$$v_* = \sqrt{\frac{\tau}{\rho}}.$$ (5.27)

When $U_T < v_*$, turbulent eddies are capable of transporting particles upward and suspension will dominate. When $U_T \gg v_*$, the particle trajectory is unaffected by turbulence and particles will be transported by saltation. Particles too large or heavy to be lifted from the surface by the wind will be moved either by impact creep or traction.

The static threshold friction speed ($v_{*t}$) is the lowest value of $v_*$ at which particles begin to move. Threshold speeds on Mars are about an order of magnitude higher than on Earth because of the thinner martian atmosphere. Particles with diameters of 115 μm have $v_{*t}$ near 1 m s$^{-1}$ under current martian atmospheric conditions (Greeley and Iverson, 1985). Larger particles require higher $v_{*t}$ to begin moving because of

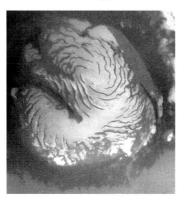

Figure 5.32 Mars' north polar cap is surrounded by an erg of saltated material. The erg appears as the dark region circling the residual polar cap. Also visible are the spiral troughs within the polar cap and layered deposits. The large entrant at left is Chasma Boreale. (MOC image MOC2-231, NASA/JPL/MSSS.)

their greater mass. Smaller particles also require higher $v_{*t}$ because of inter-particle interactions and surface roughness.

Dust is transported by suspension while sand is transported largely by saltation. These materials will be deposited when the wind velocity subsides, forming a variety of eolian depositional features. Large deposits of saltated material form an erg, or sand sea, surrounding the martian north polar cap (Figure 5.32). The largest dunefield associated with this erg lies along the 180° longitude line in a region called Olympia Planitia. Recent MOLA analysis indicates that Olympia Planitia is an extension of the polar cap (Fishbaugh and Head, 2000). The thermal inertia of the Olympia Planitia dunes indicates smaller particles than sand and likely originates from erosion of a sulfur-rich volcanic layer within the adjacent polar layered deposits (Herkenhoff and Vasavada, 1999; Byrne and Murray, 2002; Langevin *et al.*, 2005a).

Smaller deposits of saltated material form sand dunes. Crescent-shaped barchan dunes (Figure 5.33a) form when the wind blows consistently from one direction. Transverse dunes (Figure 5.33b) are the most common dune morphology on Mars. They form in regions where sand is abundant and winds are strong. Longitudinal dunes (Figure 5.33c) occur in regions with a moderate amount of sand and a prevailing wind direction. MGS analysis suggests that some dunes are still actively forming and migrating while others are currently inactive (Edgett and Malin, 2000). Smaller deposits of sand, called ripples (Figure 5.33d), have been seen at all five martian landing sites (Greeley *et al.*, 1999, 2006; Sullivan *et al.*, 2005).

Dust deposited in regions where the air flow is laminar can create thick loess deposits. The layered deposits surrounding the polar caps are composed of ice-cemented dust layers deposited during climate cycles (Sections 5.3.7 and 7.2.2).

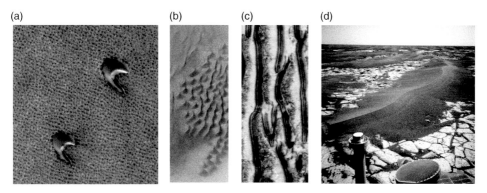

Figure 5.33 Various types of sand dunes are seen on Mars. (a) These barchan dunes are located near 73.8°N 319.2°E. (MOC image MOC2-1390.) (b) Transverse dunes are commonly seen on the floors of impact craters, such as these examples located at 51.8°S 254.5°E. (MOC image MOC2-1176.) (c) Richardson crater displays these longitudinal dunes, located at 72.4°S 180.3°E. (MOC image MOC2-1322.) (d) Ripple encountered by Opportunity during its traverse across Meridiani Planum. (Images a, b, and c are all ~3 km wide and courtesy of NASA/JPL/MSSS; image d courtesy of NASA/JPL–Caltech.)

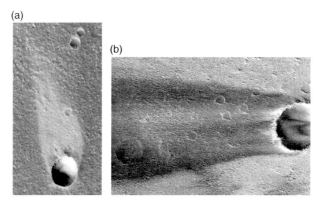

Figure 5.34 Wind streaks are often seen in the lee of martian impact craters. (a) Bright wind streaks, like this one extending from a 600-m-diameter crater near 42.0°N 234.2°E, form when dust is trapped in the lee of the crater. (MOC image MOC2-1489.) (b) Dark wind streaks form when turbulence scours dust from the lee of the crater. Crater is 688 m in diameter and located near 11.7°S 223.6°E. (MOC image MOC2-1298, NASA/JPL/MSSS.)

Bright wind streaks are another example of features resulting from dust deposition (Figure 5.34a). Bright streaks often form in the downwind direction of topographic obstacles such as impact craters. These result when winds are strong enough to remove surface dust from all regions except those protected by topography. Fine-grained dust appears brighter than coarse-grained material, so the dust remaining in

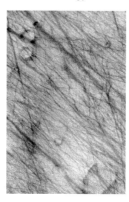

Figure 5.35 Dust devils remove fine dust from the surface, leaving behind dark streaks showing the track of the dust devil. These tracks occur south of the Hellas Basin. Image is ~3 km in width. (MOC image MOC2-1378, NASA/JPL/MSSS.)

the lee of a topographic obstacle will appear brighter than the surroundings from which the dust has been removed. Pelkey *et al.* (2001) find no strong thermal inertia difference between the streaks and the surrounding area, suggesting thicknesses of bright dust deposits between 1 μm and 3 mm. Bright wind streaks are one of the few martian geologic features seen to change over the time period of detailed investigations by orbiting spacecraft (i.e., since about 1970).

Dark wind streaks are also observed on Mars (Figure 5.34b). These features can be either depositional (composed of dark materials) or erosional due to atmospheric turbulence behind an obstacle which scours the dust from the surface. Thermal inertia of erosional dark wind streaks is distinct from the surroundings, indicating deposits of sand-sized particles thicker than a few centimeters (Pelkey *et al.*, 2001). Dark surface streaks also result when dust devils remove surface dust (Figure 5.35). Surface dust lies within the laminar regime of wind flow, making it difficult to remove (surface roughness increases the stability of surface dust). Dust devils appear to be a major mechanism in raising the surface dust above the laminar flow region for transport across the planet (Section 6.3.2).

Yardangs are tall ridges produced by wind erosion in areas of easily erodable material (Figure 5.36). They are usually oriented parallel to the prevailing wind direction, although resistant layers within yardangs can result in other orientations (Bradley *et al.*, 2002). The identification of yardangs within Mars' Medusae Fossae Formation was one of the first indications of the fine-grained nature of these deposits (Ward, 1979). Yardangs are also seen within the martian polar layered deposits (Howard, 2000).

Rocks often show the effects of eolian erosion by saltating sand particles through grooves, pits, and smoothed sides (Figure 5.37). Ventifacts are common at the martian landing sites and show that wind erosion is the major process affecting the

Figure 5.36 Yardangs are created by eolian erosion. These yardangs occur on volcanic plains west of Olympus Mons near 13.2°N 199.9°E. Image is ~3 km wide. (MOC image MOC2-1455, NASA/JPL/MSSS.)

Figure 5.37 Mars Pathfinder's Sojourner rover took this view of the pits and grooves produced by wind abrasion on Moe rock. (NASA/JPL.)

planet today. Laboratory experiments and numerical modeling indicate that kinetic energies of these sandblasting particles are about an order of magnitude higher on Mars than on Earth because of the higher wind velocities needed to initiate saltation (Bridges *et al.*, 2005).

### 5.3.6 Fluvial processes

Liquid water cannot exist on the martian surface under the current low temperature and atmospheric pressure conditions. One of the biggest surprises from the Mariner 9 investigations was the presence of channels of varying sizes on the martian surface. Channels are classified as valley networks, outflow channels, and fretted channels (Baker, 1982) (Figure 5.38). Channel morphology indicated that

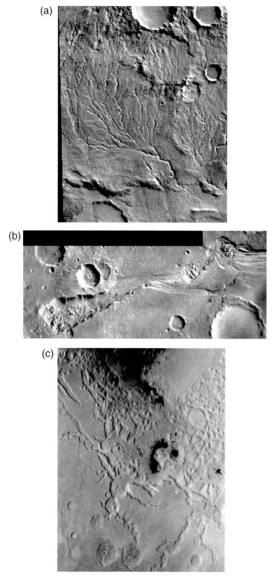

Figure 5.38 Water and/or ice has produced several types of channels on Mars. (a) Warrego Valles, located near 43°S 267°E, displays the dendritic nature characteristics of valley networks. (Viking mosaic image, NASA/JPL.) (b) Ravi Vallis (1°S 318°E), like many other outflow channels, originates from a region of chaotic terrain. The channel is about 300 km long in this image. (Lunar and Planetary Institute.) (c) Fretted terrain occurs along the hemispheric dichotomy, particularly around Arabia Terra. This view of the Nilosyrtis Mensae area (32°N 61°E) is ~248 km wide and shows the dissected nature of the fretted terrain. (MOC image MOC2-885, NASA/JPL/MSSS.)

liquid water was responsible for formation of the valley networks and outflow channels. Ice combined with tectonic activity probably created the fretted channels (Section 5.3.7).

All fluids move downhill at a velocity determined by the fluid's viscosity, the terrain, and the planet's gravity. The flow usually transports solid materials, just like the atmosphere transports small particles. Both atmospheric and fluvial transport are described by the physics of fluid motions. Thus, the physics described in Section 5.3.5 for atmospheric transport by suspension, saltation, impact creep, and traction also apply to transport of materials by fluids.

Gently flowing streams have laminar flow, which results in little erosion. Fast-flowing streams tend to be turbulent (large $R_e$) and erosive. The discharge ($Q$) of a stream is the product of the cross-sectional area ($A$) of the stream and the average flow velocity ($v$):

$$Q = vA. \tag{5.28}$$

The average flow velocity (in m s$^{-1}$) can be estimated from the Manning equation:

$$v = \frac{R^{2/3}S^{1/2}}{n} \tag{5.29}$$

where $R$ is the hydraulic radius, which is the ratio of $A$ to the extent of the wetted perimeter, $S$ is the slope of the water surface, and $n$ is the Manning roughness coefficient. Typical values of $n$ for natural streams range from 0.025 for streams on gentle slopes to 0.05 for streams with rough beds and high slopes. Irwin *et al.* (2005) scaled the Manning equation to Mars and related $Q$ to the width of the channel ($W$):

$$Q = 1.4W^{1.22}. \tag{5.30}$$

Larger discharge rates (such as occur during flooding) can move larger objects as well as transport much more material than lower discharge rates.

Morphologic variations between channels provide insights into the local topography and surface characteristics. The dendritic appearance of many martian valley networks (Figure 5.38a) indicates gentle slopes with water moving down small channels to merge with larger river networks. Steeper slopes produce channels where the tributaries are almost parallel to the main river channel. Regions with strong tectonic control (faults, fractures, or joints) produce channels with sharp angles rather than gentle meanders. Regions where channels form circular patterns indicate either domes or basins, depending on the direction of flow.

Channels can form not only from precipitation or snowmelt but also from groundwater. When the flow of groundwater stops, the overlying material can collapse into the now-dry groundwater channel, producing a sapping channel on the surface. Sapping channels often display amphitheater-shaped heads which helps distinguish them from rainfall-produced channels (Baker, 1982), although recent studies of terrestrial spring-fed rivers indicate that surface properties may dictate whether such distinguishing morphologic features occur (Lamb *et al.*, 2006).

The martian valley networks (Figure 5.38a) are integrated channel systems, with individual channels up to a few kilometers wide and lengths up to a few hundreds of kilometers (Baker, 1982). Valley networks are primarily found on Noachian units, although younger networks are seen on the flanks of some volcanoes and in Valles Marineris. These younger channel systems have been attributed to hydrothermal activity (Gulick, 1998), local precipitation (Mangold *et al.*, 2004a), snowmelt (Fassett and Head, 2006), and precipitation from volcanic gases (Dohm and Tanaka, 1999). The proposed origins of the Noachian-aged valley networks include widespread precipitation (Craddock and Howard, 2002; Howard *et al.*, 2005) or groundwater sapping (Carr, 1995; Malin and Carr, 1999). A rainfall origin of the Noachian-aged valley networks is supported by the highly integrated nature of the valley networks and drainage areas (Grant and Parker, 2002; Irwin and Howard, 2002) and presence of interior channels, which tend to form by precipitation but not by sapping (Irwin *et al.*, 2005). Estimated discharge rates of valley network systems are similar to terrestrial floods produced by rainfall ($\sim$300–3000 m$^3$ s$^{-1}$) (Irwin *et al.*, 2005). However, not all valley networks display the features indicative of rainfall, indicating that both precipitation and sapping likely have contributed to valley network formation.

Outflow channels (Figure 5.38b) are much larger than valley networks, with channel widths up to several tens of kilometers and lengths of thousands of kilometers (Baker, 1982). Outflow channels display morphologic similarities to huge catastrophic terrestrial floods, such as that which formed the Channeled Scablands of the northwestern United States (Baker, 1978). Such features include longitudinal grooves and streamlined islands (Figure 5.39). Outflow channels are Hesperian to Amazonian in age and originate from large depressions, often filled with jumbled material called chaotic terrain (Figure 5.38b). These channels form from the sudden melting of ground ice from volcanic or impact processes. The water floods across the surrounding terrain, producing the channel and its associated morphologic features. The surface material overlying the outburst area collapses into the voided region, producing chaotic terrain. Discharge rates of outflow channels are in the range of $10^6$–$10^7$ m$^3$ s$^{-1}$ (Coleman, 2005).

MOC imagery has revealed other features suggestive of fluvial activity on Mars, both in the past and perhaps even at present. Channels breach the rims of many craters in the Noachian highlands and the floors of such craters are often very flat, suggestive of paleolake deposits (Cabrol and Grin, 1999, 2001; Irwin *et al.*, 2005) (Figure 5.40a). Layered deposits are superposed on many crater and canyon floors and have been interpreted as sediments deposited in lacustrine environments

Figure 5.39 Streamlined islands occur when water was forced around topography such as impact craters. This streamlined island formed by water flow from Ares Valles being diverted around an 8-km-diameter crater. Crater is located at 20.0°N 328.7°E. (THEMIS image V06305017, NASA/JPL/ASU.)

(Malin and Edgett, 2000a) (Figure 5.40b). Mineralogic investigations of the layered deposits exposed at the Opportunity landing site in Meridiani Planum confirm that at least some of these deposits were formed as water-lain sediments (Squyres *et al.*, 2004c). Distributary fans with characteristics of delta deposits are found within craters with other indicators of paleolake activity and strongly suggest fluvial activity in the martian past (Malin and Edgett, 2003; Bhattacharya *et al.*, 2005; Fassett and Head, 2005; Lewis and Aharonson, 2006) (Figure 5.40c). Alluvial fans, formed when flowing water encounters a cliff and deposits its sediments in a fan-like pattern at the base of the cliff, have been identified on the floors of several impact craters (Moore and Howard, 2005) (Figure 5.40d). While all of these features are of Noachian age, younger gullies, some of which might still be active, are seen on steep slopes within impact craters and on canyon walls (Malin and Edgett, 2000b; Heldmann *et al.*, 2005) (Figure 5.40e). While some have proposed that the channels and gullies are formed by $CO_2$ (Hoffman, 2000; Musselwhite *et al.*, 2001) or $CH_4$ (Max and Clifford, 2001) fluids, the current conditions on Mars appear to favor $H_2O$ as the fluid that carves these features (Stewart and Nimmo, 2002). The implications of these features on the location and evolution of water on Mars will be discussed in Chapter 7.

### 5.3.7 *Polar and glacial processes*

Mars' polar regions consist of thick polar layered deposits (PLDs) composed of water ice and silicate dust capped by ice-rich polar caps. Each polar cap is composed of an extensive seasonal cap obvious from fall through spring and a permanent cap

(a)

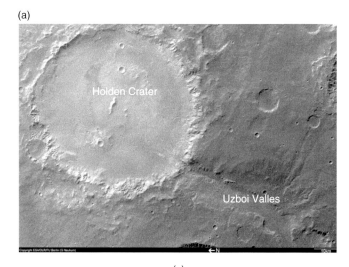

(b)                                            (c)

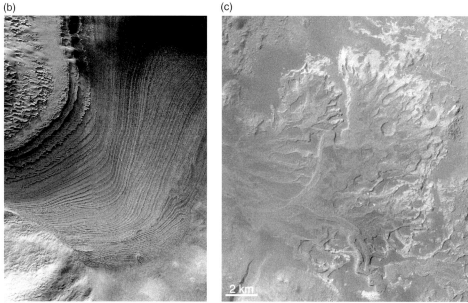

Figure 5.40 (a) A number of craters display rims breached by channels and smooth
floors. These attributes suggest that water in the channels has flooded the floors of
these craters, producing short-lived lakes. Uzboi Valles cuts the rim of 140-km-
diameter Holden Crater and may have produced a lake in the crater floor. Image
centered near 26°S 325°E. (HRSC image SEMSOYXEM4E, ESA/DLR/FU
Berlin [G. Neukum].) (b) Thick deposits of layered material are often found in
topographic depressions. This thickly layered deposit occurs within Galle Crater,
near 52.3°S 329.9°E. Image is ~4 km across. (MOC image MOC2-1494, NASA/
JPL/MSSS.) (c) This distributary fan in Eberswalde Crater (24.0°S 33.7°W)
displays features suggestive of fluvial deposition within this crater. (MOC image
MOC2-1225a, NASA/JPL/MSSS.)

(d)     (e)

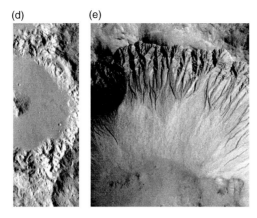

Figure 5.40 (cont.) (d) Alluvial fans descend from the rim to the floor of this ~60-km-diameter crater located at 23.0°S 74.3°E. (THEMIS image I17522002, NASA/JPL/ASU.) (e) Gullies are seen on steep slopes such as the walls of impact craters and canyons. This image, centered near 38.0°S 192.8°E, is ~3 km across. (MOC image MOC2-1292, NASA/JPL/MSSS.)

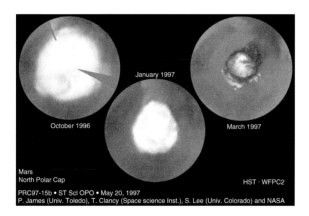

Figure 5.41 The martian polar caps display seasonal changes in size. These Hubble Space Telescope images show the change in the north polar cap from winter (left) when it is covered by the seasonal $CO_2$ ice cap until summer (right) when the permanent cap of $H_2O$ ice is visible. (University of Toledo, Space Science Institute, University of Colorado, NASA, and Space Telescope Science Institute.)

which remains through summer (Figure 5.41). Thermal and compositional characteristics of the north polar permanent cap indicate that it is composed of $H_2O$ ice (Kieffer and Titus, 2001; Langevin *et al.*, 2005b). Rheological, thermal, and spectroscopic properties indicate that the south polar permanent cap is composed of $H_2O$ ice covered by an 8-m-thick veneer of $CO_2$ ice (Nye *et al.*, 2000; Titus *et al.*,

Figure 5.42 The south polar region displays a variety of landforms, such as this "swiss-cheese" terrain caused by the sublimation of $CO_2$ ice in the spring. Image is ~1.5 km across and is located at 86.0°S 9.2°E. (MOC image MOC2-780, NASA/JPL/MSSS.)

2003; Bibring *et al.*, 2004). The north polar cap rises ~3 km above the surrounding topographically low northern plains and its center is almost coincident with the rotational pole (Zuber *et al.*, 1998). The 1100-km-diameter cap covers much of the region poleward of 80°N (Clifford *et al.*, 2000) and its volume is estimated to be between 1.1 and $2.3 \times 10^6$ km$^3$ (Zuber *et al.*, 1998; Smith *et al.*, 2001a). The cap covers most of the northern PLD, although geologic evidence suggests it was more extensive in the past (Zuber *et al.*, 1998; Fishbaugh and Head, 2000; Kolb and Tanaka, 2001).

The south polar permanent cap lies about 6 km higher than the north cap, but its relief compared to the surroundings is similar. It has a current volume between 1.2 and $2.7 \times 10^6$ km$^3$ (Smith *et al.*, 2001a). It is offset from the rotational pole, with its center near 87°S 315°E (Clifford *et al.*, 2000), although its highest point is near 87°S, 10°E (Smith *et al.*, 2001a). The 400-km-diameter south cap is much smaller in extent than its northern counterpart and does not completely cover the southern PLD, but there is evidence that it, like the north cap, was more extensive in the past (Head and Pratt, 2001; Tanaka and Kolb, 2001).

The south permanent cap displays depressions with a wide variety of shapes (Figure 5.42), unlike the north cap which shows pits, cracks, and knobs grading into the underlying PLD (Thomas *et al.*, 2000b). North polar cap features result from ablation of $H_2O$ ice from strong winds and viscous flow of the ice (Zuber *et al.*, 1998). The south permanent cap consists of two layered units deposited at different times and separated by a period of erosion. Both units are currently undergoing erosion, with the older unit eroding faster (3.6 m per Mars year) than the younger unit (2.2 m per Mars year) (Thomas *et al.*, 2005). The wide variety of features is produced by sublimation and collapse of the $CO_2$ ice veneer (Byrne and Ingersoll, 2003).

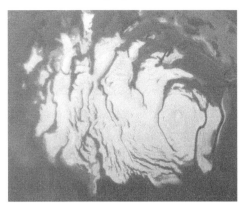

Figure 5.43 The residual south polar cap displays spiral scarps within the ice cap and underlying polar layered deposits. (MOC image MOC2-225, NASA/JPL/ MSSS.)

The north polar cap displays a series of spiral troughs into the PLD to depths up to 1 km (Figure 5.32) (Zuber *et al.*, 1998). The south polar cap displays spiral scarps (Figure 5.43). These spirals extend from the pole outward and extend into the PLD. The northern spiral troughs are separated by 20–70 km and are 5–30 km wide and up to several hundred kilometers in length (Howard, 2000). The troughs form a slight clockwise pattern, while the south polar scarps form a counterclockwise pattern. Two major formation models have been proposed for the spiral patterns. The first argues that preferential sublimation of ice from the sun-facing slopes causes the spiral pattern to originate near the edge of the polar deposits and migrate inward toward the pole (Howard *et al.*, 1982). Strong katabatic winds produced from ice-cap sublimation and deflected by the Coriolis force could enhance this erosive process since they strike the scarps at approximately right angles (Howard, 2000). The other model proposes that ice flows from the accumulation center of the cap toward the ablating edges of the polar deposits and that the spiral pattern results from the asymmetric distribution of flow centers and ice velocities (Fisher, 1993, 2000).

Large re-entrant valleys (~100 km wide and up to 2 km deep) also cut the two polar caps from the edge of the polar deposits inward to several hundred kilometers. The north pole valley, Chasma Boreale, extends halfway across the polar cap and displays steep arcuate scarps. The south pole valley, Chasma Australe, originates in an amphitheater-shaped scarp which is up to 500 m high and 20 km wide (Anguita *et al.*, 2000). Both eolian (Howard, 2000; Kolb and Tanaka, 2001) and catastrophic flood (Anguita *et al.*, 2000; Fishbaugh and Head, 2002) origins have been proposed for these chasmata.

Polar layered deposits underlying the permanent polar caps consist of horizontal layers of ice and dust (Figure 5.44) which were deposited during climate cycles

Figure 5.44 The polar layered deposits exhibit layers with differing albedos and textures. These layers are believed to represent material deposited during climate cycles. This view shows the layering exposed on a slope within the south polar region. Image is ~3 km across and is located near 86.9°S 179.5°E. (MOC image MOC2-1343, NASA/JPL/MSSS.)

driven by obliquity variations (Thomas *et al.*, 1992b) (Section 7.4.2). The northern PLDs display high thermal inertia values consistent with exposure of $H_2O$ ice while the southern PLDs display a lower thermal inertia suggesting the presence of a fine-grained lag deposit (Paige and Keegan, 1994; Paige *et al.*, 1994; Vasavada *et al.*, 2000; Putzig *et al.*, 2005). Crater SFD analysis of the PLDs indicates that the northern PLDs are younger than 0.1 Ma while the southern PLDs are about 10 Ma old (Herkenhoff and Plaut, 2000). These low crater densities suggest a resurfacing rate due to both erosion and deposition of 0.06–0.12 mm yr$^{-1}$.

Horizontal layering exposed in the PLDs ranges in thickness from 300 m down to the current resolution limits (~1.5 m/pixel with MOC) (Clifford *et al.*, 2000), although preliminary results from MRO's High Resolution Imaging Science Experiment (HiRISE) camera are revealing the limiting thickness for some layers. The albedo differences observed among layers likely result from differences in dust-to-ice content, particle size, and/or composition. While most of the layers appear to be composed of fine-grained dust, coarser particles and coherent ledges are occasionally seen (Malin and Edgett, 2001; Byrne and Murray, 2002). Unconformities between the layers and topographic relaxation of superposed impact craters indicate complex histories of deposition, ablation, subsurface creep, and erosion in both deposits (Howard *et al.*, 1982; Kolb and Tanaka, 2001; Murray *et al.*, 2001; Pathare *et al.*, 2005). Wavy layers also are seen, suggestive of internal deformation by ice flow in response to stresses caused by ice-cap loading (Fisher, 2000).

The seasonal caps, composed of $CO_2$ ice, reach their maximum extent during winter. The north polar cap is approximately symmetrical around the pole, extending to ~60°N along all longitudes. The south seasonal cap, unlike the residual cap, is centered near the geographic pole and extends to ~65°S. The seasonal cap begins forming in late summer to early fall when first thin condensate clouds and later dense clouds, called the polar hood, form over the pole. The pre-polar-hood activity begins around $L_S = 185°$ for the northern hemisphere and near $L_S = 50°$ for the southern hemisphere (Dollfus *et al.*, 1996; Wang and Ingersoll, 2002). Dust storms are common along the cap edges during this time (Benson and James, 2005). Polar hoods consist of a variety of clouds ranging from thin hazes to dense condensate clouds and contain mainly $CO_2$ ice with smaller amounts of dust and $H_2O$ ice. The seasonal cap is formed by precipitation from the polar hood. The cap becomes visible in the spring, near $L_S = 10°$ in the north and $L_S = 180°$ in the south. Albedo and emissivity variations across the seasonal cap throughout the winter result from differences in composition ($CO_2$ versus $H_2O$), grain sizes, dust abundance, and porosity (Eluszkiewicz *et al.*, 2005; Snyder Hale *et al.*, 2005). MOLA analysis indicates that seasonal precipitation adds up to 2 m to the thickness of the polar cap (Smith *et al.*, 2001b).

The spring retreat of the polar caps is non-uniform, with many frost outliers remaining after the main cap has receded. These frost outliers are typically associated with rough terrain, dunes, and crater floors where temperatures can remain low for extended periods of time, and some may be remnants from more extensive past permanent caps. Early ground-based observers noted a particularly prominent frost outlier near the south pole which was called the Mountains of Mitchell (near 72°S 40°E). This region is heavily cratered and modestly elevated relative to the surroundings. TES observations suggest that $CO_2$ frost is responsible for this region's high albedo (Kieffer *et al.*, 2000). Atmospheric circulation models propose that topography from the Hellas and Argyre impact basins enhances precipitation in this region, which combined with topography will result in frost retention (Colaprete *et al.*, 2005).

Thermal mapping of Mars reveals that cold spots are not unusual around the polar caps (Kieffer *et al.*, 2000; Kieffer and Titus, 2001; Titus *et al.*, 2001). A region centered near 77°S 160°E was detected during Viking observations to be of about the same albedo as the surroundings but remained anomalously cold. TES investigations of this "cryptic" region reveal that it is composed of fine-grained, very clear (slab) $CO_2$ ice which, due to lower thermal inertia, topography, and perhaps increased frost precipitation, remains cold while the temperature of the surroundings increases with increasing insolation (Kieffer *et al.*, 2000; Colaprete *et al.*, 2005). Dark features such as spots and fans are common in the cryptic region and represent $CO_2$ gas jetting caused by sublimation originating below the ice slab (Kieffer *et al.*, 2006). A similar cryptic zone occurs over the Olympia Planitia dunefield near the north polar cap (Kieffer and Titus, 2001). Smaller exposures of slab ice are also reported at both poles.

A variety of geological structures geomorphologically similar to terrestrial glacial features are seen in regions surrounding the present-day polar caps (Kargel and Strom, 1992; Kargel *et al.*, 1995; Head and Pratt, 2001; Ghatan and Head, 2002; Hiesinger and Head, 2002; Milkovich *et al.*, 2002). These features, together with a variety of fluvial features, suggest that the extent of the polar ice deposits has been greater in the past and underwent basal melting to achieve their present extent. Putative glacial features include sinuous ridges proposed to be eskers (ridges formed by subglacial river deposition) (Figure 5.45a), lineated terrain formed by flow of ice-rich material (Figure 5.45b), "thumbprint terrain" interpreted as terminal deposits called moraines (Figure 5.45c), and glacial-carved valleys called cirques. All of these features appear to be Hesperian to Early Amazonian in age (Kargel *et al.*, 1995; Head and Pratt, 2001; Milkovich *et al.*, 2002).

Geothermal models suggest that subsurface ice is stable up to the surface at latitudes poleward of ~50° (Clifford and Hillel, 1983), although variations in surface thermophysical properties can result in surface ice also existing in other regions (Paige, 1992; Mellon *et al.*, 1997, 2004). High-latitude regions display geologic structures that are similar to terrestrial permafrost features and which strongly correlate with subsurface $H_2O$ distribution as revealed by GRS (Kuzmin *et al.*, 2004; Mangold *et al.*, 2004b). Irregular-shaped depressions (Figure 5.46a) are similar to terrestrial thermokarst, formed by thermal degradation of ice-rich soils (Costard and Kargel, 1995). Small mounds with large summit pits (Figure 5.22) have been interpreted as pseudocraters, formed by interaction of molten lava with ground ice (Greeley and Fagents, 2001), or pingos, domes of frozen ground uplifted due to hydrostatic pressure as water freezes (Soare *et al.*, 2005). Small polygons (<500-m diameter) result from thermal contraction of volatile-rich soils (Figure 5.46b), enhanced by subsurface water filling the bounding cracks which subsequently freezes and expands during colder periods (Siebert and Kargel, 2001; Kossacki *et al.*, 2003; Mangold, 2005; van Gasselt *et al.*, 2005).

Lower-latitude features also have been interpreted as products of glacial activity from wetter climatic conditions associated with obliquity cycles (Section 7.4.2). Lucchitta (2001) has argued that longitudinal grooves and streamlined hills in the outflow channels result from warm-based glaciers. Lobate debris aprons along the flanks of the Tharsis shield volcanoes and at the bases of massifs east of Hellas appear to be ice-rich and are proposed to be the remnants of recent (Late Amazonian) glacial activity (Neukum *et al.*, 2004; Head *et al.*, 2005). Lobate debris aprons are also observed at the bases of the small gullies found along crater and canyon walls. The gullies may have formed either by seepage of groundwater (Malin and Edgett, 2000b) or from melting of a surface snowpack deposited during the last high-obliquity period (Christensen, 2003). The associated lobate debris aprons at the base of these gullies display longitudinal features and terminal ridges (Figure 5.47),

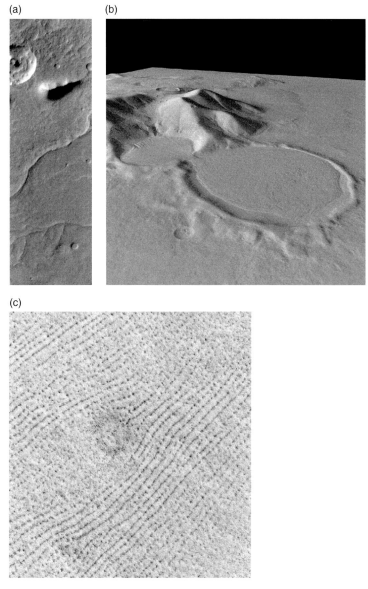

Figure 5.45 (a) Sinuous ridges, such as these seen near the southern rim of Argyre Basin, have been suggested to be eskers formed under an extensive glacier. Crater at top is ~14.3 km in diameter and located at 55.1°S 316.7°E. (THEMIS image I08553004, NASA/JPL/ASU.) (b) The floors of many channels and craters contain lineated terrain which has been interpreted as ice-rich material flowing downslope. This perspective image from HRSC shows lineated material apparently flowing from a 9-km crater into a 16-km crater. This crater is located adjacent to the Hellas Basin, near 38°S 104°E. (Image SEM1ZGRMD6E, ESA/DLR/FU Berlin [G. Neukum].) (c) Thumbprint terrain is composed of lines of small mounds. One interpretation is that these are blocks deposited in moraines of retreating glaciers. Image is ~3 km across and located at 72.4°N 107.4°E. (MOC image MOC2-513, NASA/JPL/MSSS.)

(a)                                      (b)

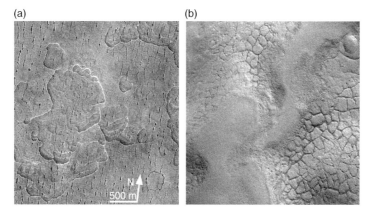

Figure 5.46 (a) Pits are seen in some regions of the northern plains where other indications of subsurface ice exist. These pits may be analogous to thermokarst pits on Earth, created by the removal of subsurface ice. These pits in Utopia Planitia are located near 44.9°N 85.3°E. (MOC image MOC2-293, NASA/JPL/MSSS.) (b) Small polygons, such as these seen near Lyot Crater at 54.6°N 33.4°E, are typically formed by freeze–thaw cycles of ice-rich soils. Image is ∼3 km across. (MOC image R1001555, NASA/JPL/MSSS.)

Figure 5.47 Glacier-like flows have been detected in several locations on Mars. This feature is located along the wall of a crater near 38.6°S 112.9°E. Image is ∼2.5 km wide. (MOC image M1800897, NASA/JPL/MSSS.)

similar to features associated with glacial flow on the Earth and may represent active debris-covered glaciers (Arfstrom and Hartmann, 2005).

Fretted terrain occurs along the dichotomy boundary primarily in the 0°–70°E longitude zone. It consists of flat-floored, steep-walled channels (fretted channels) which merge into smaller knobs inundated with plains material further to the north (Figure 5.38c). Fretted terrain likely originated from tectonic activity associated with dichotomy boundary formation and subsequently has been modified by fluvial (Carr, 1995, 1996; Carruthers and McGill, 1998; McGill, 2000), eolian (Irwin *et al.*, 2004), mass wasting (Carr, 1995), and/or glacial activity (Lucchitta, 1984; Head *et al.*, 2006a, b). The presence of lobate debris aprons, theater-headed valleys, and lineated floor fill which moves downslope through the valleys forming the fretted channels ("lineated valley fill") strongly suggests that ice-rich material continues to modify these channels at present (Head *et al.*, 2006a, b).

## 5.4 Geologic evolution of Mars

The variety of geologic processes described in Section 5.3 has led to a complicated geologic history for Mars. Impact cratering rates were highest during the Noachian, during which time destruction of pre-existing features by crater formation and deposition of ejecta blankets over surrounding regions were major processes of degradation. The record of this intense cratering period is observed in the southern highlands and has recently been revealed through a population of large craters buried under the younger northern plains deposits (Frey *et al.*, 2002; Watters *et al.*, 2006). Internal heat from accretion, differentiation, and decay of short-lived radionuclides was also higher early in martian history, leading to large-scale volcanic and tectonic activity from the Noachian to the Hesperian periods. Inter-crater plains, ridged plains, paterae, and the initial volcanism in Elysium and Tharsis began during this time and the opening of Valles Marineris was initiated. Volcanic and tectonic activity became concentrated in the Elysium and Tharsis regions during the Amazonian period as the amount of internal heat declined, with current activity focused in the region between these two provinces.

Water has played a major role in the geologic evolution of Mars. Noachian Mars had a thicker atmosphere which could have led to warmer surface conditions, allowing rainfall to occur and liquid water to exist on the planet's surface (Jakosky and Phillips, 2001) (Section 6.4). Liquid water is invoked to explain the abundance of valley network channels from this period and the increased degradation rates inferred from highly eroded Noachian surface features (Craddock and Howard, 2002). The formation of large impact basins such as Hellas and Argyre near the end of the Heavy Bombardment Period helped to remove much of the martian atmosphere, resulting in the cessation of rainfall and global liquid water reserves on the surface. Short-period localized episodes of fluvial activity associated with formation of the outflow channels were interspersed with longer periods of cold, dry conditions similar to those observed today (Baker, 2001). Obliquity variations have influenced the martian climate in recent (Amazonian) time, alternating between drier and wetter periods. Temperatures associated with the wetter periods are still debated (Section 7.4.2). If warm and wet, rainfall might occur and liquid water would be abundant in rivers, lakes, and possibly oceans. If cold and wet, snow would fall, producing glaciers even in equatorial regions. Except for these periods of high obliquity, Mars has been cold and dry in recent times, with geologic activity dominated by eolian, mass wasting, and ice-creep processes.

The mineralogic information acquired by MGS, Odyssey, and MEx provides additional constraints on the geologic evolution of Mars (Section 4.3). OMEGA observations of the temporal distribution of specific minerals subdivide the planet into three geochemical periods (Bibring *et al.*, 2006). The Phyllosian Period corresponds to the Early to Middle Noachian and indicates abundant water on Mars which produced the phyllosilicate signature of this period. Increased volcanic activity during Late Noachian and Early Hesperian periods outgassed copious amounts of sulfur, which interacted with liquid water to produce the sulfur mineralogy which characterizes the Theiikian Period. Ferric oxides dominate the mineralogy of Late Hesperian through Late Amazonian materials, indicating that liquid water has been rare in recent times on Mars except in very localized regions and for short time periods. Alteration during this Siderikian Period has been limited to frost–rock interactions and atmospheric weathering processes. Thus, the mineralogic and geologic records both suggest major changes in the processes which have contributed to the formation and modification of the martian surface over the past ~4.5 Ga.

# 6

# Atmospheric conditions and evolution

## 6.1 Characteristics of the present martian atmosphere

Eighteenth-century astronomers observed clouds on Mars, revealing the presence of the planet's atmosphere. Ground-based reflectance spectroscopy suggested that $CO_2$ was a major component of the martian atmosphere based on the large number of absorption lines in the 2–4-µm region. Mariners 4, 6, and 7 confirmed that $CO_2$ was the dominant component of the martian atmosphere (Table 6.1) but found that the current atmosphere is very thin, exerting an average pressure of only 700 Pa (7 mbar). The surface pressure can vary by up to 20% as a result of seasonal variations in the amounts of atmospheric $CO_2$ and $H_2O$ (Leovy, 2001). This thin atmosphere prevents the retention of solar heat, leading to large temperature changes between the planet's day and night sides. As noted in Section 4.4.2, a planet's equilibrium temperature ($T_{eq}$) results from the balance between incoming solar radiation (insolation) and reradiation from the planet's surface. This equilibrium temperature is given by

$$T_{eq} = \left[ \frac{(1 - A_b)L_{solar}}{16\pi r^2 \varepsilon \sigma} \right]^{1/4} = \left[ \frac{F_{solar}(1 - A_b)}{4\varepsilon \sigma r_{AU}^2} \right]^{1/4} \tag{6.1}$$

where $F_{solar}$ is the solar flux at the Earth's orbit ($F_{solar}$ = solar constant = $1.36 \times 10^3$ J s$^{-1}$ m$^{-2}$) and $r_{AU}$ is the distance of the planet from the Sun in astronomical units. In Chapter 4 we saw that Mars' average $A_b$ = 0.250. Assuming an average emissivity $\varepsilon = 1$ (applicable for rocky surfaces) and setting $r_{AU}$ = 1.52 AU, we find that Mars' $T_{eq}$ = 210K. Mars' observed average surface temperature is ~240K, indicating that it experiences a slight amount of greenhouse warming due to trapping of emitted surface heat by the $CO_2$-rich atmosphere.

The low average pressure of 700 Pa and low average temperature of 240K precludes the existence of liquid water on most of the martian surface at the present time. Emplacement of liquid $H_2O$ on Mars' surface leads to its immediate sublimation into the atmosphere or freezing into ice on the surface. Five locations have

Table 6.1 *Constituents of the martian atmosphere (by volume)*

| | |
|---|---|
| Carbon dioxide ($CO_2$) | 95.32% |
| Nitrogen ($N_2$) | 2.70% |
| Argon (Ar) | 1.60% |
| Oxygen ($O_2$) | 0.13% |
| Carbon monoxide (CO) | 0.08% |
| Water ($H_2O$) | 210 ppm |
| Nitrogen oxide (NO) | 100 ppm |
| Neon (Ne) | 2.5 ppm |
| Hydrogen–deuterium–oxygen (HDO) | 0.85 ppm |
| Krypton (Kr) | 0.3 ppm |
| Xenon (Xe) | 0.08 ppm |

been identified where pressure and temperature conditions may allow liquid water to exist for short periods of time (37 to 70 sols): between 0° and 30°N in Amazonis, Arabia, and Elysium Planitiae, and in the Hellas and Argyre impact basins (Haberle *et al.*, 2001). Whether liquid water actually occurs in these regions is uncertain, particularly since there is only weak correlation between these areas and the distribution of near-surface $H_2O$ from GRS. However, the salt-rich nature of the martian soils, as determined from the surface landers and rovers, suggests that $H_2O$ brines may exist on the surface for short periods of time.

Understanding the martian atmosphere draws largely upon our comprehension of Earth's atmospheric dynamics. The lack of oceans on Mars simplifies atmospheric dynamics compared to Earth, but complications result from events that are minor contributors to the terrestrial atmosphere. One complication is Mars' greater orbital eccentricity, which leads to larger variations in annual solar insolation than that experienced by Earth. Dust storms are common on Mars and lead to heating in the upper atmosphere due to absorption and reradiation of heat by dust particles and cooling of the lower atmospheric layers due to blockage of solar radiation reaching the surface, leading to less greenhouse warming. Regions of low thermal inertia produce "thermal continents" on Mars, which affect atmospheric circulation. Obliquity variations affect the long-term atmospheric circulation through climate change. But the major driver of atmospheric circulation on Mars is the seasonal condensation and sublimation of $CO_2$ and $H_2O$ between the atmosphere and seasonal polar caps.

## 6.2 Atmospheric physics

### 6.2.1 Barometric equation and scale height

The martian atmosphere is structured into layers which vary in composition, temperature, and the physical nature of atmospheric gases. Variations in pressure ($P$)

and density ($\rho$) as a function of altitude above the planet's surface ($z$) are given by the equation of hydrostatic equilibrium:

$$\frac{dP}{dz} = -g(z)\rho(z) \tag{6.2}$$

where $g(z)$ is the gravitational acceleration at altitude $z$. Pressure and temperature ($T$) can be related in atmospheric gases through the ideal gas law:

$$P = NkT \tag{6.3}$$

where $N$ is the number of gas particles per unit volume and $k$ is Boltzmann's constant ($k = 1.38 \times 10^{-23}$ J K$^{-1}$). Essentially, $N$ is the density of the gas divided by the mass of one gas particle. The mass of one gas particle is the product of the particle's molecular weight in atomic mass units (amu) ($\mu_a$) and the mass of one atomic mass unit ($m_{amu} = 1.660539 \times 10^{-27}$ kg):

$$N = \frac{\rho}{\mu_a m_{amu}}. \tag{6.4}$$

From Eqs. (6.3) and (6.4), we can derive a relationship for density:

$$\rho = \frac{P\mu_a m_{amu}}{kT}. \tag{6.5}$$

The variation in atmospheric pressure with altitude is given by the barometric equation:

$$P(z) = P(0)\exp\left(-\int_0^z \frac{dz}{H(z)}\right) \tag{6.6}$$

where $P(0)$ is the pressure at the surface ($z = 0$; typically corresponding to the planet's mean radius or geoid) and $P(z)$ is the pressure at some altitude $z$ above the surface. $H(z)$ is the scale height of the planetary atmosphere, which depends on the temperature ($T$), gravitational acceleration ($g$), and average molecular weight ($\mu_a$) at the altitude of interest:

$$H(z) = \frac{kT(z)}{g(z)\mu_a(z)m_{amu}}. \tag{6.7}$$

The scale height $H$ is the distance over which the pressure decreases by a factor of $1/e$. Small values of $H$ imply that the atmospheric pressure decreases rapidly with altitude. The scale height of Mars' atmosphere is about 10 km.

A similar equation can be derived for the variation in density with altitude:

$$\rho(z) = \rho(0)\exp\left(-\int_0^z \frac{dz}{H^*(z)}\right). \tag{6.8}$$

Here, $H^*(z)$ is the density scale height, given by

$$\frac{1}{H^*(z)} = \left(\frac{1}{T(z)}\right)\frac{dT(z)}{dz} + \frac{g(z)\mu_a(z)m_{amu}}{kT(z)}. \tag{6.9}$$

In regions of the atmosphere where temperature does not vary with altitude, $H^*(z)=H(z)$.

A variety of heating sources drives atmospheric motions on Mars. Solar heating is the most important driver of atmospheric circulation, caused by the absorption of visible wavelength solar photons by the surface and within atmospheric layers with moderately large optical depths (i.e., near cloud layers). The martian surface and atmospheric dust particles reradiate the visible wavelength energy they absorb, usually at infrared wavelengths which further heat the atmosphere. At higher altitudes, ultraviolet (UV) and extreme ultraviolet (EUV) radiation can break apart molecules and cause ionization of atoms and molecules, providing another source of atmospheric heating. Atmospheric dynamics are driven by the most efficient mechanism of heat transport within a particular region, leading to the layered structure of the atmosphere. The three mechanisms of transporting heat are conduction, convection, and radiation.

### 6.2.2 Conduction

As noted in Section 3.4.1, conduction transfers heat through direct collisions between atoms and molecules. Conduction is important in the upper parts of planetary atmospheres and sometimes near the surface. Fourier's equation (Eq. 3.27) and the thermal diffusion equation (Eq. 3.30) are utilized to quantitatively describe atmospheric conduction just as they are used in analysis of conduction within planetary interiors.

### 6.2.3 Convection

Convection dominates in the lower parts of the martian atmosphere. Heat is transferred between regions of different temperature through movement of the material. Convection occurs when a parcel of air in a planet's atmosphere is slightly warmer than its surroundings. The parcel expands in an attempt to re-establish pressure equilibrium. However, this expansion causes the parcel's density to decrease below that of the surroundings. The parcel then rises as it seeks a region with equivalent density. But the pressure of the surroundings decreases with height, causing the rising parcel to expand as it rises. The parcel's temperature drops as it expands. The temperature of the surroundings also decreases with height above the surface. If the temperature of the surroundings decreases sufficiently rapidly with height, the parcel remains warmer than its surroundings. Thus, the parcel continues to rise and transports heat upward. Alternately, cold air is denser than its surroundings and will descend. Atmospheres where energy transport is dominated by convection have a temperature structure which follows an adiabatic lapse rate. No heat is exchanged between the convecting parcel and the surroundings under adiabatic conditions.

The temperature structure of an atmosphere undergoing convection can be derived from the equation of hydrostatic equilibrium (Eq. 6.2) and basic thermodynamic relations. Conservation of energy within the atmosphere is obtained from the first law of thermodynamics:

$$dQ = dU + PdV. \tag{6.10}$$

The amount of heat absorbed from the surroundings by a system is $dQ$. This is related to the change in internal energy of the system ($dU$) and the work done by the system on its environment ($PdV$, where $P$ = pressure and $dV$ = change in volume). The thermal heat capacities of the atmosphere at constant volume ($C_V$) and constant pressure ($C_P$) are given by

$$\begin{aligned} C_V &= \left(\frac{dQ}{dT}\right)_V = \left(\frac{\partial U}{\partial T}\right) \\ C_P &= \left(\frac{dQ}{dT}\right)_P = \left(\frac{\partial U}{\partial T}\right)_P + P\left(\frac{\partial V}{\partial T}\right)_P. \end{aligned} \tag{6.11}$$

The specific volume ($V$) is the volume per unit mass. Density ($\rho$) is related to specific volume through

$$\rho = \left(\frac{1}{V}\right). \tag{6.12}$$

Differentiating the ideal gas law (Eq. 6.3) and substituting Eq. (6.12) for the density, we find

$$dV = \left(\frac{k}{\mu_a m_{amu}P}\right)dT - \left(\frac{kT}{\mu_a m_{amu}P^2}\right)dP. \tag{6.13}$$

The specific heat at constant pressure ($c_P$) or constant volume ($c_V$) is the heat capacity per unit mass ($m$):

$$c_P = \frac{C_P}{m} \quad \text{and} \quad c_V = \frac{C_V}{m}. \tag{6.14}$$

The difference of the thermal heat capacities (or specific heats) in an ideal gas is the gas constant ($R_{gas} = 8.31 \text{ J Mole}^{-1} \text{ K}^{-1}$):

$$C_P - C_V = m(c_P - c_V) = R_{gas} = N_A k \tag{6.15}$$

where $N_A$ is Avogadro's number, the number of particles in one mole ($=6.022 \times 10^{23}$ particles), and $k$ is Boltzmann's constant.

Assume we have an atmosphere composed of an ideal gas which is convecting. A parcel of this air moves adiabatically ($dQ = 0$). Equation (6.10) then becomes

$$dU = -PdV. \tag{6.16}$$

Using the relationships in Eqs. (6.11) and (6.14), we find that

$$c_V dT = -PdV$$
$$c_P dT = \left(\frac{1}{\rho}\right) dP. \tag{6.17}$$

The ratio of the specific heats, or thermal heat capacities, occurs often in thermodynamic applications and is indicated by the parameter $\gamma$:

$$\gamma = \frac{c_P}{c_V} = \frac{C_P}{C_V}. \tag{6.18}$$

The adiabatic lapse rate of a dry atmosphere is obtained by combining the thermodynamic equation, equation of hydrostatic equilibrium, and ratio of the specific heats:

$$\rho c_P \left(\frac{dT}{dz}\right) = -g\rho \rightarrow \frac{dT}{dz} = \frac{-g}{c_P}. \tag{6.19}$$

This is often rewritten as

$$\frac{dT}{dz} = \frac{-(\gamma - 1)}{\gamma} \left(\frac{g \mu_a m_{amu}}{k}\right). \tag{6.20}$$

Gases have specific values of $\gamma$. Monatomic gases have $\gamma = 5/3$. The value of $\gamma$ for diatomic gases is 7/5 and for polyatomic gases is 4/3. The martian atmosphere's dry adiabatic lapse rate is $4.5 \text{K km}^{-1}$.

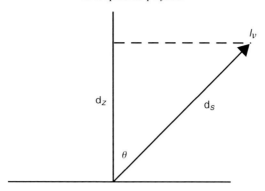

Figure 6.1 The intensity ($I_\nu$) of radiation is related to the distance it travels through an atmosphere (d$s$) and the angle at which it is emitted ($\theta$).

The dry adiabatic lapse rate is the maximum temperature gradient of a convecting atmosphere. Addition of volatiles such as water reduces this temperature gradient, giving rise to a wet adiabatic lapse rate (Section 6.3.2).

### 6.2.4 Radiation

Radiation, the absorption and re-emission of energy by atoms, is the third mechanism by which energy is transported through an atmosphere. The intensity of radiation changes as it passes through an atmosphere. This change in intensity (d$I$) depends on the radiation's initial intensity ($I$), atmospheric density ($\rho$), atmospheric thickness (d$s$), and absorption and emission properties of the atmosphere:

$$dI_\nu = j_\nu \rho ds - I_\nu \alpha_\nu \rho ds. \tag{6.21}$$

The subscript $\nu$ indicates that the intensity, emission coefficient ($j_\nu$), and absorption coefficient ($\alpha_\nu$) depend on the radiation's frequency.

Energy can be emitted at any angle and thus d$s$ can be greater than the minimum atmospheric thickness. Figure 6.1 shows the geometry of energy being reflected/scattered from a surface. From this diagram, one can see that d$s = \sec \theta$ d$z$. Let $\mu_\theta = \cos \theta$. Then Eq. (6.21) can be rewritten as

$$dI_\nu = j_\nu \rho \left(\frac{1}{\mu_\theta}\right) dz - I_\nu \alpha_\nu \rho \left(\frac{1}{\mu_\theta}\right) dz. \tag{6.22}$$

The first term on the right side of Eq. (6.22) provides information about the addition of energy by the atmosphere while the second term gives insights into how the atmosphere absorbs energy. The extinction coefficient ($\alpha_\nu$) is related to the optical depth ($\tau_\nu$) of the atmospheric gases:

$$d\tau_\nu = \alpha_\nu \rho dz. \tag{6.23}$$

The ratio of extinction and absorption coefficients is called the source function ($S_\nu$):

$$S_\nu = \frac{j_\nu}{\alpha_\nu}. \tag{6.24}$$

Applying these substitutions to Eq. (6.22) gives the equation of radiative transfer:

$$\mu_\theta \left(\frac{dI_\nu}{d\tau_\nu}\right) = S_\nu - I_\nu. \tag{6.25}$$

If $S_\nu$ does not depend on the optical depth of the atmosphere, we can integrate Eq. (6.25) to get

$$I_\nu(\tau_\nu) = S_\nu + e^{-\tau}[I_\nu(0) - S_\nu]. \tag{6.26}$$

An optically thick atmosphere has $\tau_\nu \gg 1$ and Eq. (6.26) reduces to $I_\nu = S_\nu$, indicating that the intensity of the emitted light entirely results from energy emitted from within the atmosphere. An optically thin atmosphere has $\tau_\nu \ll 1$, reducing Eq. (6.26) to $I_\nu = I_\nu(0)$. In this case, the radiation is minimally affected by its passage through the atmosphere and its intensity is defined by the incident radiation. The amount of atmosphere through which the radiation passes varies with zenith angle (angular distance of object from overhead). Thus, the change in intensity can be related to both the optical depth ($\tau_\nu$) and the zenith angle ($z$) through Beer's law:

$$I = I_0 e^{-\tau/(\cos z)}. \tag{6.27}$$

The temperature gradient can never exceed the dry adiabatic lapse rate under equilibrium conditions, so we can determine whether radiation or convection dominates within a particular region of an atmosphere by comparing these two values. Superadiabatic conditions occur when the observed thermal gradient exceeds the dry adiabatic lapse rate:

$$\left(\frac{-dT}{dz}\right)_{obs} > \frac{(\gamma - 1)}{\gamma}\left(\frac{g\mu_a m_{amu}}{k}\right) \quad \text{or} \quad \left(\frac{-dT}{dz}\right)_{obs} > \frac{-g}{c_p}. \tag{6.28}$$

Convection will drive the temperature gradient into an adiabatic lapse rate under superadiabatic conditions, so if Eq. (6.28) is satisfied we can say that convection is the primary energy transport mechanism within that part of the atmosphere.

Radiation will dominate when

$$\left(\frac{-\mathrm{d}T}{\mathrm{d}z}\right)_{\mathrm{obs}} > \frac{(\gamma - 1)}{\gamma}\left(\frac{g\mu_{\mathrm{a}}m_{\mathrm{amu}}}{k}\right). \qquad (6.29)$$

Sections of atmospheres obeying Eq. (6.29) are in radiative equilibrium.

## 6.3 Present-day martian atmosphere

### 6.3.1 Atmospheric structure

The martian atmosphere is structured in layers based on composition, temperature, isotopic characteristics of the gases, and the physical nature of atmospheric gases. This structure has been revealed by a variety of spacecraft and ground-based observations (Figure 6.2). The pressure, density, and temperature measurements made by the five lander missions (VL1, VL2, MPF, and MER) as they descended through the atmosphere provide the most detailed information about the atmospheric structure but are "snapshots" in terms of location and time. Long-term observations by orbiting spacecraft using radio occultation measurements and infrared sounding provide insights into atmospheric structure over larger regions and time periods. From these results, the martian atmosphere is subdivided into three layers: lower, middle, and upper.

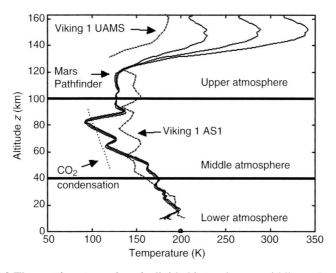

Figure 6.2 The martian atmosphere is divided into a lower, middle, and upper part based on temperature, pressure, density, and compositional variations. This figure shows the temperature profiles measured by instruments on the Mars Pathfinder and Viking 1 landers as they descended to the surface. (NASA/JPL.)

The lower atmosphere extends from the martian surface to an altitude of ~40km. Pressure and temperature decrease with increasing altitude throughout the lower atmosphere and energy transport is dominated by convection within the first 10km (Leovy, 2001). Convection ceases at night and a strong temperature inversion develops close to the surface. Temperatures and pressures within the lower atmosphere are comparable to those found in the Earth's stratosphere. The density of the lower atmosphere is largely driven by the $CO_2$ and $H_2O$ seasonal cycles associated with the sublimation and condensation of the polar caps, which in turn lead to surface pressure variations throughout the year.

Two processes heat the martian lower atmosphere. As noted in Section 6.1, the lower atmosphere experiences a slight greenhouse warming as atmospheric $CO_2$ blocks infrared radiation from the surface from escaping to space. In addition, the martian lower atmosphere contains large quantities of dust which also absorb sunlight and re-emit thermal infrared energy. Atmospheric dust is a major contributor to the heating of the lower atmosphere and must be included in circulation models of this region.

Ozone ($O_3$) may also contribute to atmospheric heating over the winter poles through absorption of solar UV and dissociation of the $O_3$ molecule. Ozone is rare over most of Mars because of the limited amount of $O_2$ and O in the atmosphere and the destruction of $O_3$ by interactions with $H_2$ (produced by photolysis of $H_2O$ vapor). The cold conditions over the winter poles reduce the amount of atmospheric $H_2O$ vapor in those locations and allow formation of some $O_3$ (Perrier *et al.*, 2006). Ozone has been detected in both the lower and middle atmospheres of Mars (Blamont and Chassefière, 1993; Novak *et al.*, 2002; Lebonnois *et al.*, 2006).

The middle atmosphere, or mesosphere, extends between 40 and ~100km above the martian surface. Middle atmosphere profiles obtained by VL1, VL2, and MPF indicate that temperatures can vary considerably with time. These temperature variations result from the near-IR absorption and emission of solar radiation by $CO_2$ and from atmospheric waves initiated in the lower atmosphere and enhanced by thermal tides between the day and night hemispheres (Schofield *et al.*, 1997).

The upper atmosphere, or thermosphere, extends beyond 110km altitude. The thermosphere is heated by solar EUV radiation, which has energies between 10 and 100 eV and wavelengths between 10 and 100 nm. Solar EUV output varies with the solar activity cycle, resulting in large variations in thermospheric temperatures above 120km altitude. Temperatures are cooler during periods of low solar activity and increase as the sunspot cycle reaches maximum. The region above 130km altitude is called the ionosphere because solar radiation ionizes the atmospheric gases above this level. Most of the electrons in Mars' ionosphere are derived from $CO_2$ and peak photoelectron densities occur on the day side (Frahm *et al.*, 2006).

The low densities and high temperatures encountered in the upper thermosphere enhance the escape of atoms from the planet. If a gas is in thermal equilibrium, the number of particles with velocities between $v$ and $v + dv$ is given by the Maxwell distribution function:

$$f(v)dv = N\left(\frac{2}{\pi}\right)^{1/2}\left(\frac{m}{kT}\right)^{3/2}v^2 e^{\frac{-mv^2}{2kT}}dv \tag{6.30}$$

where $N$ is the number density of particles, $m$ is the particle mass, $k$ is Boltzmann's constant, and $T$ is the temperature of the gas. Particle velocities typically display a Gaussian distribution, with a frequency peak at some velocity which tails off at higher velocities. In the lower levels of a planetary atmosphere, densities are high enough that collisions between the gas particles are common and the velocity distribution of these particles is driven toward the Maxwell distribution function. However, collisions are few in regions of low density and particles in the tail of the Maxwellian velocity distribution can have velocities exceeding the escape velocity. The altitude above which particles can escape to space is called the exobase and occurs above ~130–150km altitude on Mars (Mantas and Hanson, 1979).

### 6.3.2 Clouds and dust storms

Ground-based observers noted that albedo markings on Mars are often obscured and correctly attributed these changes to clouds within the martian atmosphere. Martian clouds are divided into yellow clouds, white clouds, and polar hoods (discussed in Section 5.3.7). Hazes, which are optically thin while clouds are optically thick, are often seen along the terminator, particularly along the sunrise limb, and have been observed from the surface landers (Figure 6.3). These result from vapor condensation during the low nighttime temperatures. Early morning fog also is seen in low-lying areas such as the Valles Marineris canyon and inside impact craters (Figure 6.4).

Yellow clouds have been observed since 1877 and are now recognized as dust storms. Dust is prevalent across the martian surface because of the lack of liquid water and the range of geologic processes which have eroded surface rocks. Longitudinal differences in atmospheric pressure and temperature can produce strong winds capable of producing dust storms. Dust devil activity, which is more common in the southern hemisphere and extends from late spring to early fall (Fisher *et al.*, 2005; Whelley and Greeley, 2006), helps lift dust into the atmosphere (Basu *et al.*, 2004; Kahre *et al.*, 2006) (Figure 6.5). Martian dust storms can be local, regional, or

Figure 6.3 Early morning clouds and hazes are common sights from the landers. This image shows thin cirrus clouds over the Mars Pathfinder landing site. (NASA/JPL/Imager for Mars Pathfinder team.)

Figure 6.4 Fog is often seen in topographic depression in the early morning. This MOC image shows a 36-km-diameter crater, located at 66.4°S 151.4°E, filled with fog. (MOC image R0700964, NASA/JPL/MSSS.)

Figure 6.5 Dust devils help to lift dust from the surface into the martian atmosphere. These dust devils were captured by Spirit's camera as they traveled across the floor of Gusev Crater. (NASA/JPL/Texas A&M University.)

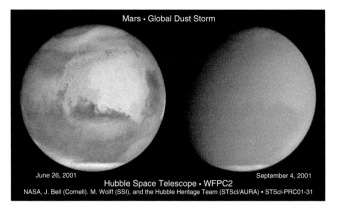

Figure 6.6 Global dust storms can arise quickly on Mars, as demonstrated in these two images taken 2.5 months apart by the Hubble Space Telescope. Albedo features are easily discerned in the image from 26 June 2001, but are completely covered by the dust storm in the image from 4 September 2001. (Image STScI-PRC01-31, NASA/Cornell/SSI/STScI/AURA.)

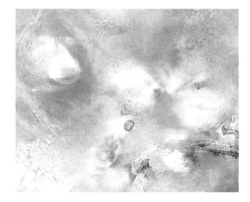

Figure 6.7 White clouds can be seen over the Tharsis volcanoes and western Valles Marineris in this MOC regional view. Orographic clouds commonly occur near the tall martian volcanoes. (MOC image MOC2-144, NASA/JPL/MSSS.)

global in extent, depending on the atmospheric conditions. Global dust storms (Figure 6.6) typically occur near perihelion when it is summer in the southern hemisphere. The stronger daytime heating associated with Mars' proximity to the Sun produces the strong winds and dust devil activity which initiate dust storms. Cooling of the surface under the dust storm leads to further temperature gradients and more wind, causing the dust storm to expand. Under the right conditions, this mechanism can continue until the entire planet is engulfed in dust, whereupon the surface temperature variations diminish, the winds decrease, and the dust storm ends.

White clouds and hazes have been observed since the seventeenth century and are primarily composed of $H_2O$, although some $CO_2$ clouds have also been detected (Figure 6.7). The white clouds clearly show on UV and blue filter images of Mars. Many of the white clouds are orographic clouds, produced as the atmosphere is forced by topography to higher altitudes where the lower temperatures allow condensation of the $H_2O$ vapor. Ground-based observers in the seventeenth century often reported the presence of a W-shaped cloud in the western hemisphere of Mars. Today we know that the W-shaped cloud is an orographic cloud associated with the Tharsis volcanoes.

Convecting air parcels will rise at the dry adiabatic lapse rate until they reach a level where the temperature equals the condensation temperature of one of the gas components, at which point clouds will form. Air is saturated when the abundance of a condensable gas is at its maximum vapor partial pressure (the amount of atmospheric pressure contributed by the vapor). Evaporation (or sublimation) is balanced by condensation in saturated air. Liquid droplets will condense when more vapor is added to saturated air, resulting in the formation of clouds. The amount of vapor contained in the atmosphere and how close that atmosphere is to saturation is measured by the relative humidity. Relative humidity (RH) compares the partial pressure of the vapor ($p_v$) to that of saturated air ($p_s$):

$$\mathrm{RH} = 100 \left( \frac{p_v}{p_s} \right). \tag{6.31}$$

The value of RH is close to 100% in clouds. The saturation vapor pressure at a particular temperature is given by the Clausius–Clapeyron equation:

$$p_s = C_L e^{\frac{-L_s}{R_{gas}T}} \tag{6.32}$$

where $C_L$ is a constant specific to the type of gas condensing and $L_s$ is the latent heat of condensation for saturated air. The release of this latent heat affects the temperature gradient and thus the lapse rate in this region of the atmosphere. The specific heats of the condensing gas become

$$c_V = -P \frac{dV}{dT} - L_s \frac{dw}{dT} \tag{6.33}$$

$$c_P = \frac{1}{\rho} \frac{dP}{dT} - L_s \frac{dw}{dT}. \tag{6.34}$$

The mass of water vapor which condenses out per gram of air is given by $w$. The $c_p$ in the dry adiabatic lapse rate (Eq. 6.20) must be adjusted using Eq. (6.34) to include the heating effects from the condensation:

$$\frac{dT}{dz} = \frac{-g}{c_P + L_s\left(\frac{dw}{dT}\right)}.$$ (6.35)

This thermal gradient is called the wet adiabatic lapse rate. The wet adiabatic lapse rate can never exceed the dry adiabatic lapse rate, therefore the latter is often used as an upper limit to the thermal gradient of a convecting atmosphere.

### 6.3.3 Winds

Gradients in pressure and temperature resulting from solar heating produce winds in an attempt to reduce these gradients. Atmospheric pressure and temperature gradients result from three major factors: seasonal changes, dust storms, and diurnal variations. The seasonal changes result from condensation and sublimation of $CO_2$ and $H_2O$ from the polar caps. As noted above, dust storms can enhance the temperature gradient leading to stronger winds. The diurnal variations result from the temperature differences between the day and night sides of the planet as well as the presence of passing storm systems. Diurnal wind directions often change throughout the day (Figure 6.8). Examples of the winds triggered by these processes are Hadley circulation, thermal tides, and condensation flows.

If a planet's rotation axis lies perpendicular to the ecliptic plane, the equator receives more solar energy than other latitudes. Convection in the atmosphere causes warmer air to rise and flow toward regions with lower temperature and pressure. Thus the warm air rises over the equator, cools, and sinks back to the surface near the poles. The flow then returns to the equator along the surface. For slowly rotating or

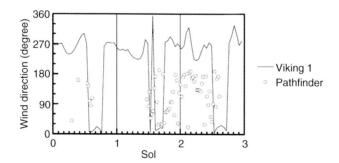

Figure 6.8 Wind direction varied considerably throughout a martian sol at both the Mars Pathfinder and Viking 1 landing sites. Wind direction is the azimuth angle, measured from north (0°). Wind direction at the Pathfinder site typically changed from south-southwest winds in the morning to north-northeast in the afternoon. (MPF image SS009, NASA/JPL.)

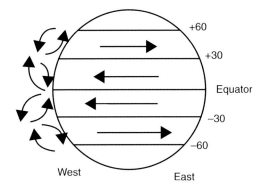

Figure 6.9 Differences in solar insolation with latitude combined with a planet's rotation produce Hadley cells which drive atmospheric circulation. Hadley cell circulation produces easterly winds (winds from east to west) within the equatorial zone and westerlies in the 30° to 60° latitude zone in each hemisphere. Arrows on the left show the vertical circulation within each latitude zone, with warm air rising over the equator and cooler air sinking near ±30° latitude.

non-rotating planets, there are two convection cells, one in the northern hemisphere and one in the south. These convection cells are called Hadley cells.

Venus is an example of a planet with one Hadley cell per hemisphere. The faster rotation of Mars causes the north–south winds (meridional winds) to be deflected and the Hadley cells split into three Hadley cells per hemisphere due to conservation of angular momentum (Figure 6.9). This results in east–west (zonal) winds occurring along the surface, with easterlies (winds from east toward west) dominating near the equator and westerlies (winds from west toward east) occurring in the mid-latitude zones. The velocity gradient of these winds can be estimated from the thermal wind equation:

$$\frac{du}{dz} \simeq \frac{1}{H}\frac{\partial u}{\partial \ln P} = \frac{g}{2R\Omega(\sin\phi)}\frac{\partial\theta}{\partial\phi} \tag{6.36}$$

where $H$ is the pressure scale height (Eq. 7.7) at altitude $z$, $u$ is the zonal wind velocity, $P$ is the pressure and $R$ is the planet's radius. The angular velocity ($\Omega$) of the atmosphere in the rotating frame of the planet depends on the latitude ($\phi$) of the atmospheric parcel because of the Coriolis force. The potential temperature, $\theta$, is given by:

$$\theta = T\left(\frac{P_0}{P}\right)^{R_{gas}/c_P}. \tag{6.37}$$

Equation (6.36) predicts zonal wind velocities near 85 m s$^{-1}$ near the top of the Hadley circulation and ~10 m s$^{-1}$ near the surface (Read and Lewis, 2004).

The pattern displayed in Figure 6.9 only results when the planet's rotation axis is perpendicular to its orbital plane. Any tilt of the rotation axis causes the Hadley cell circulation to be displaced from the equator and produces seasonally changing weather patterns. The large eccentricity of Mars' orbit also leads to large time-averaged differences between the polar regions. Thus, while Mars' wind patterns are dominated by Hadley cell circulation, complications to this simplified view occur because of its rotation rate, obliquity, and orbital eccentricity as well as the presence of polar caps and thermal continents.

Mars' thin atmosphere is not very efficient at retaining the daytime heating, leading to dramatic temperature decreases at night. The large temperature difference between day and night hemispheres causes air flow from the warmer day side to the cooler night side. Winds produced by this temperature gradient are called thermal tides. Thermal tides are centered in the equatorial regions but extend to mid-latitudes. The thermal gradient and thus the efficiency of thermal tides can be estimated by comparing the solar heat input, $F_{in}$, with the heat capacity of the atmosphere. Solar heat input per day is related to the surface area exposed to solar radiation, the amount of the solar radiation absorbed, and the length of the day ($t_d$):

$$F_{in} = \pi R^2 (1 - A_0) \left( \frac{F_{solar}}{r_{AU}^2} \right) t_d \qquad (6.38)$$

where $R$ is the radius of the planet, $A_0$ is the geometric albedo, $F_{solar}$ is the solar constant, and $r_{AU}$ is the distance of the planet from the Sun in astronomical units. The amount of heat, $Q$, necessary to raise the atmospheric temperature by some temperature change, $\Delta T$, is related to the heat capacity of the atmosphere per unit area and the total area of the atmosphere:

$$Q = c_P \left( \frac{P_0}{g} \right) \left( 4\pi R^2 \Delta T \right) \qquad (6.39)$$

where $P_0/g$ is the mass of the atmosphere per unit area of the surface. If all of the solar heat absorbed by the atmosphere is used to raise the atmospheric temperature by $\Delta T$, then $F_{in} = Q$. The fractional increase in the temperature becomes

$$\frac{\Delta T}{T} = \frac{[F_{solar}(1 - A_0)gt_d]}{4P_0 c_P r_{AU}^2 T}. \qquad (6.40)$$

Planets with tenuous atmospheres, like Mars, display much larger fractional temperature changes than planets with thick atmospheres. For Mars $\Delta T/T$ is approximately 20%, compared to ~0.4% for Venus.

Condensation flow winds result from the condensation of $CO_2$ over the fall/winter pole and its sublimation over the spring/summer pole. Atmospheric pressure

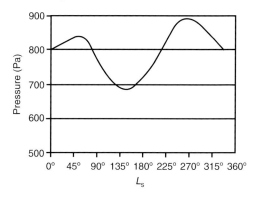

Figure 6.10 Atmospheric pressure varies on an annual cycle because of the condensation flow between the polar caps. This figure shows the pressure variation averaged over three years at the Viking 1 landing site near latitude 22.5°N. Highest atmospheric pressures occurred near northern winter ($L_s=270°$) with lowest pressures occurring near the end of northern summer.

increases as the $CO_2$ moves into the fall/winter hemisphere and decreases as summer approaches and the $CO_2$ migrates to the opposite hemisphere. The combination of these polar sinks/sources of $CO_2$ and Mars' high orbital eccentricity produce a 20% variation in surface pressure throughout the year (Figure 6.10).

The condensation/sublimation of $CO_2$ is a major driver of Mars' atmospheric circulation and is called the $CO_2$ seasonal cycle (James *et al.*, 1992). The $CO_2$ cycle is weakly linked with the dust cycle since the upwardly directed wind associated with sublimation will transfer dust from the ice cap to the atmosphere while the downward wind at the condensing polar cap will deposit dust (Kahn *et al.*, 1992; James *et al.*, 2005).

The amount of solar insolation decreases during the fall and reduces to zero at the poles during the winter. The amount of energy lost at the top of the winter atmosphere due to thermal radiation is not balanced by the energy produced by atmospheric convection. The temperature can therefore drop to the condensation temperature of $CO_2$ (148 K), causing atmospheric $CO_2$ to condense and form the seasonal polar cap. Latent heat from the condensation of $CO_2$ is the major atmospheric energy source during polar night.

The $CO_2$ seasonal cycle involves up to 30% of the martian atmospheric mass. Atmospheric circulation is strongly affected as this mass sublimes from the spring/summer pole and is transferred to the fall/winter pole for condensation. This condensation flow adds ~0.5 m s$^{-1}$ to the planet's Hadley-circulation-produced meridional winds (Read and Lewis, 2004).

Another contributor to the condensation flow winds is the $H_2O$ seasonal cycle (Jakosky and Haberle, 1992; Houben *et al.*, 1997; Richardson and Wilson, 2002).

This cycle involves the exchange of $H_2O$ between atmospheric and non-atmospheric reservoirs. The non-atmospheric reservoirs include the seasonal and permanent polar caps, adsorbed water in the regolith, and surface or near-surface ice. Most of the atmospheric transport involves $H_2O$ vapor, although the white condensate clouds also contribute. As with $CO_2$, the abundance of atmospheric $H_2O$ vapor varies with latitude and season, ranging over a factor of two.

The dominant processes which affect the abundance of $H_2O$ vapor in the atmosphere are sublimation from the seasonal and permanent polar caps, condensation on the fall/winter polar cap, desorption of $H_2O$ from regolith grains due to seasonal temperature changes, and diffusion of $H_2O$ vapor from the regolith into the atmosphere. The permanent caps are the dominant contributor to atmospheric $H_2O$ by pumping $H_2O$ into the atmosphere in the spring/summer and removing it in the fall/winter. This creates an imbalance in the amount of $H_2O$ between the hemispheres, which can be counteracted by the regolith absorbing or releasing $H_2O$. Equilibrium between the atmosphere and regolith is achieved within a few years.

### 6.3.4 Atmospheric circulation

The general circulation of the martian atmosphere is driven by zonal and meridional winds, the Coriolis force, planetary waves, and seasonal condensation flow. Zonal winds and meridional circulation are produced by solar insolation through Hadley circulation, although the seasonal condensation flow also contributes to the meridional component.

Winds follow a curved path from high to low pressure areas because of the planet's rotation. For prograde rotating bodies like Mars, winds are deflected to the right in the northern hemisphere and to the left in the southern hemisphere. This deflection of wind patterns because of the planet's rotation is the Coriolis effect and the fictitious force causing the wind to curve is the Coriolis force. The Coriolis force ($F_C$) is an expression of the conservation of angular momentum:

$$\vec{F}_C = 2\vec{\Omega} \times \vec{v} \tag{6.41}$$

where $\Omega$ is the angular velocity of the planet and $v$ is the velocity of the atmosphere. The magnitude of $F_C$ depends on the latitude $\phi$, being greatest near the poles and negligible near the equator:

$$F_C = 2\Omega v(\sin \phi). \tag{6.42}$$

Atmospheric waves are another major contributor to Mars' atmospheric circulation. Stationary waves, also called Rossby waves, remain fixed in the rotating reference frame of the planet and are associated with strong eastward jets (the jet stream

is a familiar example). They propagate vertically in response to topographic variations or from heating over thermal continents. They tend to occur at high latitudes
because of the strength of the Coriolis force and, while strongest in the winter
hemisphere, occur at all seasons. On Mars, they influence the stability of the
atmosphere and are a major contributor to the distribution of heat from the equatorial
to the polar regions. MGS's TES and Radio Occultation investigations were the first
to unambiguously detect stationary waves in the martian atmosphere (Banfield *et al.*,
2000, 2003; Hinson *et al.*, 2001, 2003; Fukuhara and Imamura, 2005). Radio
occultation results suggest that stationary waves dominate at altitudes below ~75 km
(Cahoy *et al.*, 2006).

Traveling planetary waves are produced by temperature and pressure (baroclinic)
variations and are commonly associated with weather fronts. Clouds associated with
traveling waves were detected in Mariner 9 and Viking data (Conrath, 1981;
Murphy *et al.*, 1990), but the most detailed view of traveling waves has been
obtained from MGS analysis (Hinson and Wilson, 2002; Wilson *et al.*, 2002).
Banfield *et al.* (2004) analyzed two years of MGS TES data to determine that
traveling waves are strongest in late northern fall and early northern winter, with
much weaker waves detected in the southern fall and winter seasons. They also
found strong annual repeatability in traveling waves.

The atmospheric circulation is modeled using Mars Global Circulation Models
(MGCMs), which grew out of terrestrial GCMs beginning in 1969. MGCMs primarily model the circulation occurring within the lower atmosphere, although some
have been extended to the middle and upper atmosphere. MGCMs parameterize the
physical processes affecting atmospheric circulation and include contributions from
Hadley circulation, Coriolis effect, atmospheric dust, radiative heating and cooling,
clouds, convection, turbulence, waves, drag forces from interaction with surface
roughness (i.e., the planetary boundary layer), and the seasonal condensation/
sublimation flow (Read and Lewis, 2004). The four major MGCMs have been
developed at NASA Ames Research Center (Pollack *et al.*, 1990, 1993; Barnes
*et al.*, 1993, 1996; Haberle *et al.*, 1993; Murphy *et al.*, 1995; Joshi *et al.*, 1997), the
French Laboratoire de Météorologie Dynamique (Hourdin *et al.*, 1995; Forget *et al.*,
1999), Oxford University (Joshi *et al.*, 1995; Collins *et al.*, 1996; Lewis *et al.*, 1997;
Newman *et al.*, 2002a, b), and Princeton University (Wilson, 1997; Richardson and
Wilson, 2002; Richardson *et al.*, 2002). These MGCMs have become increasingly
robust and successful in modeling the observed martian atmospheric circulation.

GCMs provide insights into the global circulation patterns but computational
limitations prevent them from providing much detail. In recent years mesoscale
models have been developed which provide detailed studies of small-scale processes on a regional scale (Rafkin *et al.*, 2001; Toigo and Richardson, 2002).
Mesoscale models are particularly useful when investigating processes such as dust

lifting (Toigo *et al.*, 2003) and conditions at landing sites (Tyler *et al.*, 2002; Kass *et al.*, 2003; Toigo and Richardson, 2003).

### 6.3.5 Present-day martian climate

Observations from orbiters and surface landers combined with the numerical modeling from the MGCMs provide a reasonably detailed understanding of the present-day martian atmosphere. Daytime heating results in convection which can lift dust off the surface through dust devil activity from late spring through early summer (Hinson *et al.*, 1999). Convection can extend through an entire scale height ($\sim$10km). The temperature gradient in the convective region approaches that of the dry adiabatic lapse rate of $\sim$4.3Kkm$^{-1}$. At night, convection ceases and strong temperature inversions develop. Nighttime temperatures can drop low enough to cause condensation of atmospheric water vapor, producing the hazes and fog commonly seen in the early morning hours. Thermal tides dominate the circulation at altitudes above $\sim$75km, particularly in the tropics, while planetary waves dominate at lower altitudes (Cahoy *et al.*, 2006).

Eastward zonal winds dominate in the mid-latitude zones and increase in intensity vertically up to about 30km during all seasons except for mid-summer when winds emanate from the west. Lower latitude winds are dominated by westerlies all year.

Hadley circulation is the dominant contributor to the meridional winds although the seasonal condensation flow between the poles cannot be ignored on Mars. During the equinoxes a symmetric pattern of Hadley circulation develops on either side of the equator, but near the solstices the ascending branch is displaced near 30° latitude in the summer hemisphere and the descending branch occurs near 60° latitude in the winter hemisphere (Haberle *et al.*, 1993). Heating variations induced by Mars' large orbital eccentricity enhance these patterns, making the northern winter Hadley circulation stronger.

Planetary waves disrupt the general Hadley circulation pattern in mid to high latitudes. Eastward-propagating traveling planetary waves are produced by variations in atmospheric temperature and/or pressure (baroclinic and/or barotropic instabilities). The number of wave lobes along a particular latitude band is called the wave number of the planetary wave – martian traveling waves typically have zonal wave numbers between 1 and 3 during the winter and spring (Hollingsworth *et al.*, 1996). Weather systems are commonly seen along the edges of the polar caps (Figure 6.11).

Stationary waves develop through the interaction of the traveling planetary waves with topography or thermal continents. Stationary waves typically have zonal wave numbers of 1 or 2 (Conrath, 1981). As the atmosphere flows over the massive topographic features on Mars, internal gravity waves are generated which further affect the stability and heating of the atmospheric layers. As the atmosphere ascends

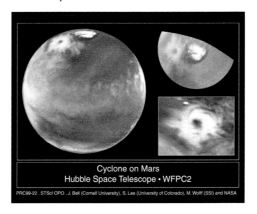

Cyclone on Mars
Hubble Space Telescope • WFPC2

PRC99-22 . STScI OPO . J. Bell (Cornell University), S. Lee (University of Colorado), M. Wolff (SSI) and NASA

Figure 6.11 Storm systems often occur near the polar caps. This Hubble Space Telescope image shows one such storm occurring near Mars' north polar cap. (Image PRC99-22, Cornell University/University of Colorado CO/SSI/NASA/ STScI.)

over the large martian volcanoes, the drop in temperature with altitude causes condensation and the formation of orographic clouds.

Wind direction is affected by the seasonal drift of the Hadley circulation and the strength of the condensation flow. The tropical Hadley circulation zone produces winds blowing from the northeast in the northern hemisphere and from the southeast in the southern hemisphere. However, the seasonal shift in the location of the ascending and descending branches of the Hadley circulation means that mid-latitudes can experience these trade winds during the winter, and this is generally consistent with wind directions inferred from wind-streak analysis (Thomas *et al.*, 2003). The cross-equatorial Hadley circulation during the solstices produces a westerly jet in the subtropics ($\sim 15°$–$30°$) within a scale height of the surface, with wind speeds near 33 m s$^{-1}$ at an altitude of $\sim 2$ km (Hinson *et al.*, 1999). This is much higher than the surface wind speeds measured by landers, which are typically in the 5 to 10 m s$^{-1}$ range (e.g., Smith *et al.*, 1997). However, higher wind speeds are expected to occur along topographic slopes (Joshi *et al.*, 1995).

Daytime convection and its associated dust devil activity help to lift surface dust into the atmosphere (Basu *et al.*, 2004; Kahre *et al.*, 2006) which can facilitate the production of dust storms. MOC image analysis finds that dust storms in the size range of $10^4$–$10^6$ km$^2$ tend to occur in mid-latitude zones along the edges of the seasonal polar cap (Wang *et al.*, 2005) and at lower latitudes during northern winter when Mars is at perihelion (Cantor *et al.*, 2001; James and Cantor, 2002). Dust is carried by the ascending branch of the Hadley circulation and distributed over a wide area, heating the atmosphere and causing enhancement of the associated winds so that more dust is lifted. When conditions are favorable, this process can produce the global-wide dust storms (Zurek and Martin, 1993).

## 6.4 Evolution of the martian atmosphere

The present-day martian atmosphere precludes the existence of liquid water on the surface for extended periods, but the geologic and mineralogic evidences suggest that this has not always been the situation. The abundance of valley network systems, the highly degraded nature of impact craters and other geologic features, and the existence of phyllosilicates all suggest that the Noachian period was much wetter than the conditions prevailing since that time. The fivefold enhancement of deuterium to hydrogen in the martian atmosphere relative to Earth also suggests that Mars had a thicker atmosphere in the past (Owen *et al.*, 1988; Krasnopolsky, 2000; Jakosky and Phillips, 2001). Volcanism, including formation of the highland paterae and Tharsis, was the likely source of the atmospheric $CO_2$, but volcanoes also release substantial quantities of water vapor. Accretion, differentiation, and decay of short-lived radioactive elements would have led to higher heat flow and increased volcanism during the Noachian. The $CO_2$ and $H_2O$ outgassed by these volcanoes would produce a thick early atmosphere.

Both $CO_2$ and $H_2O$ are greenhouse gases which would increase the surface temperature and, combined with the higher surface pressures provided by a thicker atmosphere, might allow rainfall and the presence of liquid water. However, stellar models indicate that 4 Ga ago the Sun was only 25–30% as luminous as it is today (Newman and Rood, 1977). This decreased luminosity translates to an effective temperature of only $\sim$196 K, requiring 77 K of greenhouse warming (Haberle, 1998). Numerical modeling suggested that a $CO_2$-rich atmosphere producing 5 bars ($5\times10^5$ Pa) of surface pressure would bring the surface temperature to 273 K with this lower luminosity Sun (Pollack *et al.*, 1987). However, $CO_2$ will condense under these higher pressure scenarios, producing clouds and releasing latent heat higher in the atmosphere (Kasting, 1991). Both of these mechanisms result in cooling of the surface below the freezing point of water. Possible solutions to this conundrum include a scattered greenhouse effect from $CO_2$ ice clouds (Forget and Pierrehumbert, 1997), greenhouse warming by other gases (Kasting, 1991; Squyres and Kasting, 1994; Sagan and Chyba, 1997; Yung *et al.*, 1997), a more massive and thus brighter early Sun (Whitmire *et al.*, 1995), and wetter microclimates produced by impacts (Segura *et al.*, 2002).

Valley network formation and the high degradation rates recorded by geologic features cease $\sim$3.9 Ga ago, suggesting that the thicker atmosphere disappeared by the end of the Noachian. The combination of three mechanisms probably led to the erosion of the martian atmosphere. Prior to $\sim$4 Ga ago, Mars' magnetic field was operating, which would have provided protection from the solar wind. Once the magnetic field ceased operation, solar wind particles could strike the martian atmosphere and strip it from the planet through sputtering and collisions (Jakosky *et al.*, 1994). The timing also coincides with the period when the cataclysmic Late Heavy Bombardment is postulated to have occurred throughout the inner Solar System (Gomes *et al.*, 2005), which would have produced many of the large impact basins. As the bolides passed through the atmosphere, friction would heat the atmospheric gases and allow many of them to escape. This impact erosion would have been particularly efficient at Mars given the planet's small size and gravitational pull (Melosh and Vickery, 1989). Solar-wind stripping and impact erosion combined with normal Jeans' escape of lighter atoms from above the exobase could have thinned the martian atmosphere to its current levels in a relatively short period of time (Brain and Jakosky, 1998; Jakosky and Phillips, 2001)

Short-lived excursions of the martian climate from its present state have occurred since the major event at the end of the Noachian. These excursions are indicated by the geologic (e.g., Baker, 2001) and geochemical (Romanek *et al.*, 1994; Watson *et al.*, 1994; Jakosky and Jones, 1997; Squyres *et al.*, 2004c) evidence of liquid water on the martian surface in the post-Noachian period. These recent climate fluctuations likely result from changes in Mars' obliquity (Section 7.4.2).

# 7

# History of water on Mars

Our understanding of the role that water has played in Mars' history has evolved dramatically over the past 40 years (Carr, 1996). Early spacecraft investigations revealed a cold, dry world with little evidence of liquid water ever having played a dominant role. Mariner 9 and Viking observations of the surface geology led to the paradigm of an early wet and warm Mars which transitioned into its present cold, dry state by the end of the Noachian. Data from MGS, Odyssey, and MEx reveal that water has played an important role up to recent times (Figure 7.1), although whether this water has been primarily in the liquid or ice form is still debated. While the atmosphere and polar caps are the most obvious locations of $H_2O$ today, the majority of Mars' water resides in the subsurface, primarily in the form of ice. The distribution of these subsurface $H_2O$ reservoirs is only now being revealed through instruments such as Odyssey's GRS and the ground-penetrating radars on MEx and MRO.

## 7.1 Origin of water on Mars

Outgassing associated with impact crater formation and volcanism is the primary source of martian $H_2O$ found in the atmosphere, polar caps, and subsurface (Pepin, 1991; McSween *et al.*, 2001). This indicates that $H_2O$ was incorporated into the crust and interior of Mars. The two possible mechanisms for incorporating this water into the planet are through volatile-rich planetesimals which were accreted into Mars or emplacement of a volatile-rich veneer through later delivery by cometary and asteroidal impacts.

Solar nebula condensation models indicate that temperatures inward of ~2.5 to 3 AU were too hot for $H_2O$ to be present (Cassen, 1994; Drouart *et al.*, 1999). Planetesimals formed within this region would be expected to be volatile-poor, although incorporation of small amounts of hydrated minerals might have occurred (Drake and Righter, 2002). The formation of Jupiter and Saturn perturbed volatile-rich

| Geologic period | Mineralogic period | Geologic processes | Mineralogy | Atmosphere | Other |
|---|---|---|---|---|---|
| Early Noachian | Phyllosian | Valley networks, high erosion, alluvial fans, paleolakes, distributary fans | Phyllosilicates | Dense; wetter | Magnetic dynamo, high impact rates (LHB), Initiation of Tharsis, Widespread volcanism |
| Middle Noachian | | | | | |
| Late Noachian | Theiikian | | Sulfates | | |
| Early Hesperian | | Low erosion rates, Outflow channels | | Thin; dry with wetter periods interspersed | No magnetic field; low rates of geologic activity |
| Late Hesperian | Siderikian | Northern ocean? | Anhydrous ferric oxides | | |
| Early Amazonian | | Cold-base glaciers, Gullies | | | Obliquity cycles |
| Middle Amazonian | | | | | |
| Late Amazonian | | | | | |

Figure 7.1 Geologic processes, dominant mineralogy, and atmospheric conditions have varied over Mars' history. This chart summarizes the major processes and characteristics of Mars from its earliest history (Early Noachian) to the present (Late Amazonian).

planetesimals from the region of those planets into the inner Solar System, providing some volatile-rich material to be incorporated into material forming the terrestrial planets (Morbidelli *et al.*, 2000). However, the amount of water incorporated into Mars during accretion must have been relatively small because SNC meteorite analysis indicates that the martian mantle is quite dry (Section 2.3).

The low water concentrations in the martian mantle can be reconciled with the surface evidence of substantial water by having Mars form from volatile-poor planetesimals and later acquire a volatile-rich veneer (Carr and Wänke, 1992). Much of the terrestrial $H_2O$ inventory was probably delivered through impacts of large volatile-rich planetesimals during the late stages of accretion (Morbidelli *et al.*, 2000; Robert, 2001). However, the small size of Mars indicates that it did not experience as many of these planetesimal impacts (Chambers and Wetherill, 1998; Chambers, 2004). Mars' water must therefore originate primarily from impacts of smaller comets and asteroids (Lunine *et al.*, 2003).

The ratio of deuterium (D) to hydrogen (H) provides constraints on the origin of water on Earth and Mars. Deuterium is the isotope of hydrogen with one neutron. Deuterium is rarer than hydrogen but its greater mass causes it to remain behind as hydrogen escapes during fractionation processes. The D/H ratio for the bulk Earth is $149 \times 10^{-6}$, which is substantially higher than the solar ratio deduced from solar wind implanted in lunar soils ($20 \times 10^{-6}$) (Geiss and Gloeckler, 1998) or the expected protosolar nebula value of $\sim 80 \times 10^{-6}$ (Robert *et al.*, 2000). This higher terrestrial value indicates that water on Earth is not derived from the solar nebula or from solar wind implantation. Clay minerals in carbonaceous chondrites have D/H ratios similar to the terrestrial ratio, although organic material in these meteorites has much higher D/H (Robert *et al.*, 2000). The heterogeneity of D/H throughout the Solar System suggests that the water in our Solar System originated by interactions of the solar nebula with deuterium-enriched interstellar ice. Turbulent mixing within the solar nebula caused the $H_2O$ to condense at different locations and at different times, producing the observed isotopic heterogeneities.

Values of D/H have been measured in comets Halley (Balsiger *et al.*, 1995), Hyakutake (Bockelée-Morvan *et al.*, 1998), and Hale-Bopp (Meier *et al.*, 1998). All three show much higher D/H ratios than terrestrial water, indicating that cometary impacts contribute no more than $\sim 10\%$ of the water in the volatile-rich veneer of the Earth. Dynamical models of cometary impacts also indicate that they would be a minor contributor to inner Solar System $H_2O$ inventories (Morbidelli *et al.*, 2000).

Migration of the giant planets $\sim 0.7$ Ga after Solar System formation perturbed volatile-rich objects in the outer asteroid belt, providing a mechanism for the delivery of $H_2O$ into the inner Solar System (Petit *et al.*, 2001; Gomes *et al.*, 2005; Strom *et al.*, 2005). The similarity of D/H in carbonaceous chondrites and the Earth complements the dynamical arguments that asteroids are the primary source of the terrestrial $H_2O$ budget. Mars, being closer to the asteroid belt than the Earth, might be expected to have acquired even higher amounts of $H_2O$ from asteroid impacts. Analysis of D/H in martian meteorites indicates that Mars is enriched in deuterium compared to Earth, probably as a result of mixing between magmatic water and a deuterium-enriched atmospheric component (Watson *et al.*, 1994; Leshin, 2000; Lunine *et al.*, 2003). The higher concentration of deuterium in the martian atmosphere is expected if Mars has lost much of its original atmosphere (Donahue, 1995) (Section 6.4).

Water delivered by asteroid collisions should have provided Mars with enough water to produce a global ocean between 600 and 2700 m deep (Lunine *et al.*, 2003). The amount of water needed to have produced all the fluvial and glacial features on Mars would produce a global ocean ~400–500 m deep (Carr, 1996). Analysis of incompatible trace elements in the martian meteorites suggest that only ~50% of the $H_2O$ incorporated into the planet's interior has been degassed (Norman, 1999; Lunine *et al.*, 2003). Summing the outgassed and non-outgassed amounts leads to an estimate that the amount of water contained by Mars is equivalent to a ~1000-m-deep global ocean. Thus, asteroid collisions can easily deliver the water inventory estimated for Mars.

## 7.2 Water versus other volatiles

The current climatic conditions preclude the existence of large quantities of liquid water on the martian surface, but the presence of geologically young channels and gullies indicates that some substance has flowed recently. Water is the favored fluid because it is one of the most common molecules in the Solar System, is expected to have been delivered in sufficient quantities to Mars, and can produce the observed characteristics of the channels and gullies. However, other volatiles also have been proposed to explain the Hesperian- and Amazonian-aged channels and gullies.

Hoffman (2000) argued that the outflow channels could be produced by $CO_2$ density flows rather than liquid water. Geothermal models of the pressure and temperature conditions in the near-surface region of Mars suggest that condensed $CO_2$ could be stable at depths of ~100 m in the mid-latitudes of the planet and liquid $CO_2$ could be present within the equatorial substrate. Hoffman's model proposes that this subsurface $CO_2$ exists as extensive clathrate deposits. Landslides or geothermal heating of $CO_2$ saturated regolith would release the gas, producing a density flow which could erode the outflow channels. Alternately, Max and Clifford (2001) proposed that the outflow channels were produced by dissociation of subsurface methane clathrates. However, the amount of $CH_4$ necessary to support such debris flows is much larger than current estimates of $CH_4$ in the martian regolith (Section 8.4).

A liquid $CO_2$ origin also has been proposed to explain the small gullies located on steep slopes such as the interior walls of impact craters in mid-latitude regions (Musselwhite *et al.*, 2001). Musselwhite *et al.* (2001) note that the triple point of $CO_2$ corresponds to a depth of $\sim$100 m in the martian regolith, which is consistent with the depth of the gully seepage locations. Their model proposes that atmospheric $CO_2$ diffuses into the regolith to a depth of 100 m. During the winter, the low temperatures cause a plug of $CO_2$ ice to form on the cliff-face and grow in thickness inward. Temperatures and pressures increase as this plug grows, leading to formation of liquid $CO_2$ interior to the ice plug. As summer approaches and temperatures increase, the ice plug begins to thin, allowing the liquid $CO_2$ eventually to break through and flow out of the cliff-face. The liquid $CO_2$ vaporizes, producing a debris flow incorporating rocks and clathrate ice which erodes the gully.

Stewart and Nimmo (2002) modeled the timescales and depths for atmospheric $CO_2$ diffusion into the martian regolith. They found that $CO_2$ can only diffuse a few meters into the regolith and that it would take longer than one winter for $CO_2$ to fill the regolith pore spaces. They also considered formation of $CO_2$ ice, liquid, and clathrates under denser atmospheric conditions such as those which existed early in Mars' history. Their results indicate that such deposits would not persist to the present day. The amount of $CO_2$ necessary to support the debris flow creating a single outflow channel is much greater than the plausible $CO_2$ content of the current regolith.

The modeling of Stewart and Nimmo (2002) argues that gullies and outflow channels probably formed by the rapid melting and transport of liquid water. The detection of $H_2O$ within the upper meter of the martian regolith by GRS and the mineralogic evidence of hydrated minerals on the surface of Mars from OMEGA, TES, THEMIS, and the MER investigations also suggest that $H_2O$ has been the dominant volatile affecting the surface geology and mineralogy throughout martian history.

## 7.3 Water on early Mars

Mariner 9 and Viking images revealed the valley network and outflow channel systems, leading to the conclusion that liquid water had been responsible for formation of these features. The valley networks are concentrated on ancient Noachian terrains, with a few on the flanks of younger volcanoes (Baker, 1982; Gulick and Baker, 1990) (Figure 7.2). Rainfall and groundwater sapping are proposed formation mechanisms for the Noachian-aged valley networks, and geomorphic analysis suggests that both mechanisms contributed (Section 5.3.6).

Figure 7.2 Valley networks are primarily concentrated on the ancient Noachian surfaces, but these valleys also occur on the flanks of younger volcanoes. These valley network systems occur on the flanks of Alba Patera. Image is centered near 46 °N 247 °E and is ~19 km wide. (THEMIS image V11912007, NASA/ JPL/ASU.)

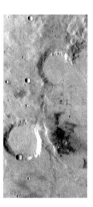

Figure 7.3 Impact craters from the Noachian period appear highly degraded, suggesting that erosion rates were higher early in martian history. The elevated rims of degraded craters have been stripped away, the walls are heavily eroded, and the floors are flat and typically filled by later deposits. Geologic processes sometimes have removed part of the crater rim, as seen in the crater near the top of this image. The lower crater is ~13.5 km in diameter and located at 31.0 °S 0.5 °E. (THEMIS image I08951002, NASA/JPL/ASU.)

Mariner 9 and Viking images also revealed that older impact craters on Noachian terrains appeared more degraded than younger craters on the planet (Figure 7.3). This suggested that the Noachian was characterized by higher erosion rates than those that have operated in the Hesperian and Amazonian. Topographic profiles of the degraded craters approximated the profiles expected for erosion by liquid water and suggested that rainfall was common on early Mars (Craddock and Howard,

2002; Forsberg-Taylor *et al.*, 2004). MGS and Odyssey images revealing alluvial fans (Moore and Howard, 2005), sedimentary deposits (Malin and Edgett, 2000a), distributary fans (Malin and Edgett, 2003; Bhattacharya *et al.*, 2005; Lewis and Aharonson, 2006), and possible paleolake deposits (Cabrol and Grin, 2001; Irwin *et al.*, 2005) further support the idea that liquid water was common on the surface of Noachian Mars.

Geologic features of fluvial origin provide one of the evidences that the atmosphere of Noachian Mars was denser, producing warmer and wetter surface conditions. Isotopic ratios of heavy noble gases in Mars' atmosphere, including D/H, indicate a thicker atmosphere than at present (Owen and Bar-Nun, 1995). Water delivered by asteroid impacts after differentiation was incorporated into the martian lithosphere and later released by outgassing associated with impacts and volcanism (Pepin, 1991), primarily during the formation of the Tharsis volcanic province during the Noachian. Impact erosion by the LHB cataclysm combined with solar wind erosion of the atmosphere after the magnetic dynamo ceased could remove up to 99% of the atmospheric gases, transitioning Mars into the thinner atmospheric conditions which exist to the present (Jakosky and Phillips, 2001).

OMEGA's detection of hydrated minerals within Noachian materials (Bibring *et al.*, 2005; Poulet *et al.*, 2005) provides mineralogic evidence that is consistent with the geologic and isotopic arguments for water on early Mars. Phyllosilicates are found in association with eroded outcrops and in dark deposits which likely resulted from erosion of ancient clay-rich materials. These locations indicate that the phyllosilicates formed early in martian history through interaction with surface water, were subsequently buried to protect them from weathering processes operating throughout most of martian history, and have only recently been exposed. The detection of phyllosilicates led Bibring *et al.* (2006) to characterize the early to middle Noachian as the phyllosian period during which liquid water was abundant on the martian surface.

The exhumation of ancient materials which have been buried for much of martian history is suggested not only by mineralogic evidence but also from geologic analysis. Several locations across the planet show craters which are being exposed as overlying layers are removed (Figure 7.4) and the relatively fresh nature of the Hesperian-aged deposits investigated by Opportunity suggests that Meridiani Planum has recently been exhumed from under younger deposits. This suggests that future missions can analyze ancient deposits which have not undergone substantial alteration by recent geologic processes.

The late Noachian was a period of transition from the mineralogic viewpoint. Increased volcanism, forming not only the volcanic constructs in Tharsis and Elysium but also the flood basalts of the ridged plains regions, released considerable

Figure 7.4 Arabia Terra is one of the locations where buried craters are reappearing as overlying younger materials are removed. Such craters initially appear as infilled circles, such as these craters, located near 28.6°N 42.5°E. The largest crater (upper right) is ∼1.4km in diameter. (MOC image R0901004, NASA/JPL/MSSS.)

amounts of sulfur into the martian atmosphere. The sulfur interacted with water to produce sulfate minerals. This theiikian period (Bibring *et al.*, 2006) extended through the early Hesperian. The highly acidic and saline aqueous conditions detected by Opportunity within Meridiani Planum date from this period (Squyres *et al.*, 2004c) and are probably representative of the aqueous environments that existed during the late Noachian to early Hesperian. By the end of the early Hesperian, surface water became rare as the martian atmosphere thinned and climatic conditions cooled. Recent martian history is characterized by anhydrous conditions except for local regions (Bibring *et al.*, 2006).

## 7.4  Water in the post-Noachian period

The post-Viking view of Mars was of a planet where liquid water had been abundant during the planet's early history but whose atmosphere had thinned by the end of the Noachian, transitioning into the cold dry conditions which exist to the present. Although evidence of recent aqueous activity was apparent in the Hesperian-aged outflow channels, these features were proposed to result from cataclysmic release of subsurface water through volcanic or impact processes. Any changes to the atmosphere were short-lived and localized and did not radically affect the climate of Mars.

### 7.4.1  Martian oceans

This view began to change around 1990 when a few scientists argued that a multitude of enigmatic features in the northern plains could be explained as evidences of past oceans within the northern lowlands (Figure 7.5). Parker *et al.* (1989)

Figure 7.5 Geologic analysis combined with MOLA topography and roughness have led to the hypothesis of a large ocean covering much of the martian northern plains in the Hesperian to Amazonian period. The extent of Oceanus Borealis is shown in lighter tone in this image superposed on a MOLA-derived shaded relief map. The Tharsis volcanoes and Valles Marineris are seen near the bottom. (NASA/JPL/GSFC/MOLA team/Brown University.)

cited the gradational nature of the dichotomy boundary in western Deuteronilus Mensae as evidence of sedimentary deposition and proposed that this deposition occurred in a sea produced during outflow channel flooding of the northern plains. Later work identified two possible shorelines which were largely continuous around the entire northern plains, leading to the proposal of one or more oceans forming in this area during Hesperian to Amazonian times (Parker *et al.*, 1993). Baker *et al.* (1991) and Kargel and Strom (1992) mapped the distribution of possible flood, permafrost, and periglacial features across Mars and argued that the hydrologic cycle associated with outflow channel formation could produce short-lived northern oceans and southern glaciers into the Amazonian period.

Models of the hydrologic cycle producing oceans in the northern plains invoke high internal heat flow to release water from subsurface ice reservoirs or aquifers. Clifford and Parker (2001) argue that this process first produced a northern plains ocean shortly after formation of the dichotomy boundary. Recharge of the deep subsurface aquifer by polar basal melting permitted subsequent outbursts through the outflow channels into more recent times, producing episodic flooding of the northern lowlands. As the geothermal flux declines with time and the permafrost layer thickens, outbursts become rarer and ocean episodes diminish. An alternate model (Baker *et al.*, 1991; Gulick *et al.*, 1997; Baker, 2001) also invokes internal heat to trigger the outflow channel flooding through melting of subsurface ground ice and $CO_2$ clathrate (Figure 7.6). The release of the $CO_2$ from the clathrates as well as volcanism leads to greenhouse warming and rainfall. The northern ocean (Oceanus Borealis) produced by the outflow channel flooding begins to evaporate

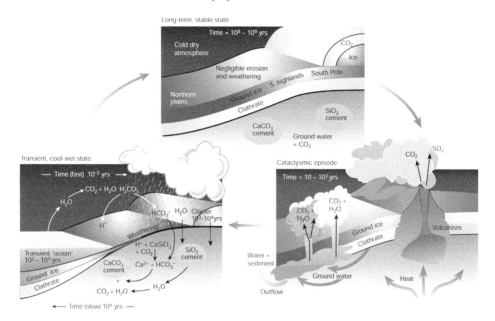

Figure 7.6 One possible scenario for how northern oceans could be produced on Mars is shown in this diagram. This model (Baker, 2001) suggests long-term cold conditions which are interspersed by short-lived cataclysmic episodes of volatile release. (Reprinted by permission from Macmillan Publishers Ltd: *Nature*, Baker [2001], Copyright 2001.)

and the moisture is transferred to the southern hemisphere where it is precipitated as snow to produce glaciers. Carbon dioxide is removed from the atmosphere by dissolving in the ocean water and through silicate weathering. Within $10^3$ to $10^5$ years, cold temperatures return and the $CO_2$ is sequestered into $CO_2$ clathrates underneath the ground ice resulting from infiltration of the ocean water. Another pulse of high heat flow can cause the cycle to repeat.

MOLA analysis reveals that one of Parker *et al.*'s (1993) proposed shorelines corresponds with an equipotential surface continuous around the northern plains (Head *et al.*, 1998, 1999). Roughness studies of the northern plains indicate they are extremely smooth, similar to sediment-covered ocean floors on Earth (Figure 4.3) (Kreslavsky and Head, 2000; Smith *et al.*, 2001a). However, MOC imagery does not reveal features which would be expected along shorelines (Figure 7.7) (Malin and Edgett, 1999) and the mineralogic data do not reveal evidence of hydrated minerals within the proposed ocean boundaries (Bibring *et al.*, 2005, 2006). Olivine, which alters quickly in the presence of water, has been detected in Hesperian-aged outcrops and argues against a recent hydrologic cycle involving liquid water on the surface. An ice-covered ocean might alleviate some of these concerns,

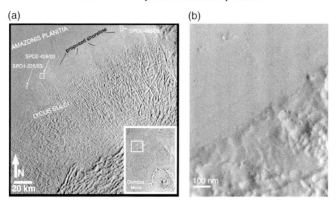

Figure 7.7 MOC has imaged the locations of proposed shorelines from Oceanus Borealis. (a) This context image shows the location of one of the proposed shorelines northwest of Olympus Mons. (b) This MOC image corresponds to the square identified as SPO2-428/03 on the context image. Although this image lies along the proposed shoreline, no evidence of such a feature is seen. (MOC images MOC2-180a and MOC2-180c, NASA/JPL/MSSS.)

but at the present time there is still considerable debate as to whether Hesperian-to-Amazonian-aged oceans have existed on Mars.

### 7.4.2 Obliquity cycles and climate change

Long-term climate change likely results from perturbations induced by the planets and the Sun's non-spherical shape. These perturbations result in variations in Mars' orbital parameters (including eccentricity, inclination, and time of perihelion) and obliquity (Ward, 1992) and are analogous to the Milankovitch cycles which contribute to ice ages on Earth. Although the obliquity and orbital motion are strongly chaotic and numerical solutions are only accurate for the past 60 Ma, numerical solutions have been extended over longer time periods to estimate the range of values (Laskar and Robutel, 1993; Touma and Wisdom, 1993; Laskar *et al.*, 2004). Those models indicate that eccentricity varies between $\sim$0 and 0.12 with a most probable value of 0.068 and orbital inclination varies between $0°$ and $8°$. Obliquity can vary from almost $0°$ to over $80°$ with a most probable value of $41.8°$. These ranges are much greater than the variations experienced by the Earth, primarily because Phobos and Deimos are too small to provide a stabilizing influence on Mars. The obliquity and orbital variations lead to insolation changes which can influence the martian climate (Figure 7.8).

The poles receive more solar insolation as the obliquity increases, increasing their temperatures and causing sublimation of the $CO_2$ and $H_2O$ ice in these regions. The addition of these gases increases the density of the atmosphere and the

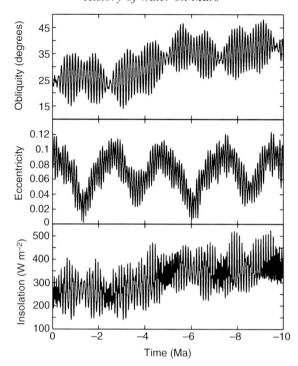

Figure 7.8 Gravitational perturbations from the other planets cause Mars to experience variations in its obliquity (top), eccentricity (middle), and solar insolation (bottom). Numerical modeling shows how these parameters have varied over the past 10 Ma. (Reprinted by permission from Macmillan Publishers Ltd: *Nature*, Laskar *et al.* [2002], Copyright 2002.)

accompanying surface pressure. Since $H_2O$ and $CO_2$ are greenhouse gases, one expects that the surface temperature will rise and could allow liquid water to exist on the surface. Depending on the combination of orbital and obliquity values, liquid water can occur in regions that range from very localized, such as occurs today (Hecht, 2002), to a large fraction of the planet (Richardson and Mischna, 2005). Ice dominates over liquid on the surface at higher obliquities because of thermal blanketing produced by $CO_2$ condensation in thick atmospheres and the greater extent of the winter polar cap (Haberle *et al.*, 2003; Mischna *et al.*, 2003). The stability zone for $H_2O$ ice moves towards the equator as the obliquity increases and is found preferentially at higher elevations and in regions of high thermal inertia.

The obliquity cycle has a period of about 2.5 Ma while eccentricity variations occur on timescales of ~1.7 Ma. Mars is currently moving from a higher to lower obliquity period and eccentricity is moving toward a high (Figure 7.8). Analysis of MOC, THEMIS, and HRSC imagery reveals features in near-equatorial zones which are interpreted as cold-based glaciers produced during the last high-obliquity

Figure 7.9 Flow lines within this deposit have been interpreted as a cold-based glacier emplaced during the last high-obliquity period. This deposit is located on the northwestern flank of Olympus Mons, near 22.5°N 222.0°E. Image is approximately 33 km wide. (THEMIS image I16967012, NASA/JPL/ASU.)

period (Head and Marchant, 2003; Neukum *et al.*, 2004; Head *et al.*, 2005, 2006a) (Figure 7.9). Tongue-like deposits on crater walls at higher latitudes (Figure 5.47), the possible source of melting water to produce gullies (Christensen, 2003), may be debris-covered glaciers from the last high-obliquity period (Arfstrom and Hartmann, 2005). Ice-rich mantling deposits occur in the ~30°–60° latitude zone in both hemispheres (Mustard *et al.*, 2001) and may have been deposited when obliquity last reached ~30°–35° (Head *et al.*, 2003a). Layering in the polar layered deposits (Figure 5.44) may record depositional cycles of ice and dust associated with the obliquity cycles (Jakosky *et al.*, 1993; Laskar *et al.*, 2002).

## 7.5 Present-day stability and distribution of $H_2O$

Current atmospheric conditions on Mars produce a triple point pressure of water several orders of magnitude higher than the water vapor partial pressure. This prevents liquid water from being stable for extended periods of time on the surface. Short-lived episodes of recent flooding associated with volcanic activity have been recorded at Cerberus Fossae (Burr *et al.*, 2002; Head *et al.*, 2003b) and a nearby feature has been interpreted as a frozen sea (Murray *et al.*, 2005) (Figure 7.10), although this feature is more commonly mapped as a basaltic lava flow. Localized regions may allow liquid water to form and flow for short periods of time under current conditions if ice deposits receive sufficient insolation (Haberle *et al.*, 2001; Hecht, 2002). However, $H_2O$ on present-day Mars is primarily in the form of ice or vapor. Condensing all of the water vapor in the martian atmosphere onto the surface would create a layer ~$10^{-5}$ m thick across the entire planet. Water in the polar caps

Figure 7.10 This unusual terrain near the Cerberus Fossae fissure resembles pack ice and has been proposed to be a frozen sea. The sharpness and number of impact craters have led others to interpret these features as resulting from basaltic lava flows. (Image SEMYVLYEM4E, ESA/DLR/FU Berlin [G. Neukum].)

and PLD would produce a global layer with a maximum thickness of 29.6 m. However, the amount of water needed to produce all the fluvially created features on Mars is estimated to form a global layer 400 m thick (Carr, 1996). Clearly there must be another reservoir of martian water – that reservoir is probably underground.

### 7.5.1 Models of subsurface $H_2O$ distribution

The layered ejecta morphologies surrounding young impact craters, central pit structures, source regions of outflow channels, valley network characteristics indicative of sapping, and various permafrost-type features (Section 5.3) suggest that much of Mars' water inventory exists in subsurface reservoirs. Hydrologic models of present-day Mars suggest that $H_2O$ ice can be stable at the surface in contact with the atmosphere at latitudes poleward of $\sim40°$ because of the low temperatures while ice at lower latitudes is stable at depths $>1-2$ m (Clifford and Hillel, 1983; Mellon and Jakosky, 1993, 1995). Models indicate that the thickness of the ice-rich cryosphere has nominal values of $\sim2.5$ km in the equatorial region to $\sim6.5$ km at the poles (Clifford, 1993). The estimated geothermal flux of 30 mW m$^{-2}$ could result in liquid groundwater aquifers under the cryosphere (Figure 7.11).

Aquifer water can migrate toward the colder cryosphere where it freezes at the base. The cryosphere is therefore expected to thicken over time. The aquifer can further be depleted when its pressure overcomes the hydraulic head of the overlying material and fluid is released in outflow channel events. However, the aquifer is replenished through pressure-induced melting at the base of the polar caps. This fluid infiltrates into the underlying material and recharges the aquifer (Clifford,

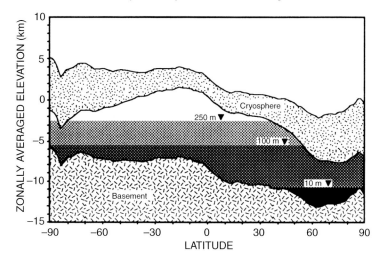

Figure 7.11 Hydrologic models of the martian substrate, using estimated values of internal heat flux and material physical properties, suggest that ice-rich (cryosphere) and water-rich regions can exist below the martian surface. The thicknesses of these regions are latitude-dependent, based on topography and solar insolation. (Reprinted by permission from the American Geophysical Union, Clifford [1993], Copyright 1993.)

1987, 1993; Clifford and Parker, 2001). Thus, large H$_2$O reservoirs are expected to exist to the present day within the upper few kilometers of the martian substrate.

## 7.5.2 Direct detection of subsurface H$_2$O

The hypothesis of substantial quantities of H$_2$O in the subsurface has been based entirely on indirect evidence from geology until recently. The neutron detectors (Neutron Spectrometer and HEND) on Odyssey's GRS instrument suite provide the first direct information about H$_2$O ice within the upper meter of the martian surface (Boynton *et al.*, 2002; Feldman *et al.*, 2002, 2004a; Mitrofanov *et al.*, 2002). The instruments detect neutrons produced by cosmic ray interactions with surface materials. These neutrons can pass through the surface before escaping to space. Hydrogen (typically bound in the form of water for terrestrial planets) is a good absorber of epithermal (energies between 0.4 eV and 0.7 MeV) and fast (energies between 0.7 and 1.6 MeV) neutrons while CO$_2$ is a poor absorber of epithermal and thermal energies (<0.4 eV) (Feldman *et al.*, 2002, 2004a). Thus, H$_2$O distribution maps can be produced by comparing the fluxes of thermal, epithermal, and fast neutrons.

Figure 4.10 shows the water-equivalent hydrogen content of martian soils derived from analysis of epithermal neutron flux (Feldman *et al.*, 2004a). Polar regions have high water abundances extending to $\sim\pm60°$ latitude, consistent with the expected distribution of $H_2O$ based on ground-ice stability models (Mellon *et al.*, 2004) and the distribution of permafrost-related geologic features (Kuzmin *et al.*, 2004). Two equatorial regions, Arabia and the highlands region south of the dichotomy boundary between Elysium and Tharsis, contain about 8% water-equivalent hydrogen and could result either from hydrated minerals (Feldman *et al.*, 2004b; Fialips *et al.*, 2005) or remnant ice emplaced by atmospheric processes (Jakosky *et al.*, 2005).

The distribution of $H_2O$ in the near-surface materials as indicated by epithermal neutron analysis has provided ground-truth for models of the stability of near-surface ice on Mars. The zones of ice stability are dictated by the balance between vapor diffusion of the ice and condensation of ice from the atmosphere. These conditions result from heating of the surface by solar insolation. The depths to which temperatures are high enough to cause soil desiccation vary on diurnal, seasonal, and long-term (i.e., obliquity-driven) cycles and reach maximum values of 1–2 m for the long-term cycles (Mellon and Jakosky, 1995). Below this desiccated zone permanent ice deposits can occur, consistent with the impact crater record which shows that layered ejecta morphologies and central pits have formed at least from Hesperian through Amazonian times (Barlow, 2004, 2006). A two-layer model consisting of a dry soil layer overlying a deep icy layer accurately reproduces most features of the epithermal neutron data in the polar regions (Boynton *et al.*, 2002; Feldman *et al.*, 2004a). Models of ice stability zones generally correlate with the epithermal neutron indicators of subsurface ice distribution (Mellon *et al.*, 2004). Longitudinal variations are largely produced by thermal inertia differences (Schorghofer and Aharonson, 2005) and other slight discrepancies between observation and model may be due to rocks and dust (Sizemore and Mellon, 2006). The correlation between observed $H_2O$ and that predicted from the ice stability models indicates that the ground ice is in equilibrium with the current martian atmosphere.

The GRS instruments provide information about $H_2O$ only within the upper meter of the martian surface, yet hydrologic models (Clifford, 1993) and impact crater morphologies (Barlow and Perez, 2003; Barlow, 2005) suggest that volatiles extend to depths of a few kilometers. Direct investigation of volatile compositions and phases at these depths are being conducted by the ground-penetrating radars on Mars Express (MARSIS: Picardi *et al.*, 2005) and Mars Reconnaissance Orbiter (SHARAD: Seu *et al.*, 2004). Those results will provide important constraints for understanding the planet's hydrologic cycle, the vertical distribution of present-day volatiles, and the role of subsurface volatiles in the formation of specific geologic features.

# 8

# Search for life

The possibility of current (extant) or past (extinct) life on Mars has been of interest for centuries. Percival Lowell's early ideas of intelligent life forms who engineered an elaborate canal system have been disproved, but the question of whether microbial life exists or has existed on Mars remains an area of intense debate. The question of martian life is one of the driving forces behind the continuing exploration of the planet.

## 8.1 Martian conditions relevant to biology

Surface conditions on present-day Mars are extremely inhospitable to terrestrial life forms (Clark, 1998). The thin $CO_2$ atmosphere is inefficient at retaining daytime solar heating, resulting in a temperature range of 130–250 K with an average temperature of 240 K. No evidence of biologic replication has been observed in terrestrial organisms at temperatures $<253$ K (Beaty *et al.*, 2006). The low temperatures and low atmospheric pressure at the surface prevent liquid water from existing for extended periods of time. Terrestrial life forms require water for survival, so the lack of liquid water on Mars is a major deterrent to life on the surface.

The thin martian atmosphere and the lack of a present-day magnetic field allow harmful radiation to penetrate to the surface (except perhaps in regions where rocks retain strong remnant magnetization [Alves and Baptista, 2004]). Odyssey's MARIE instrument (Badhwar, 2004) measured the radiation environment above the martian atmosphere until October 2003 when particles from a large solar flare caused it to cease functioning. Although it only operated for 19 months, MARIE determined that the daily radiation dose from galactic cosmic rays was between 18 and 24 millirad (1 rad$=0.01$ J kg$^{-1}$). Protons emitted during solar flare events were recorded to have typical doses of 1–2 rad and the largest events provided doses $\geq 7$ rad (Cleghorn *et al.*, 2004). While these are relatively small doses (the average cosmic ray exposure on Earth is 30–45 millirad yr$^{-1}$), the higher dosage solar particle events require shielding to protect cells from damage.

Ultraviolet (UV) radiation is another major concern on the surface of Mars. The martian atmosphere contains very little ozone except over the winter pole, therefore allowing UV to reach the surface. Models suggest that UV fluxes are up to three orders of magnitude more damaging to DNA on the martian surface than on Earth (Cockell *et al.*, 2000), and laboratory experiments conducted under martian UV conditions found rapid inactivation of seven species of bacillus (Schuerger *et al.*, 2006). However, UV conditions on present-day Mars may not be drastically different from those on Archean Earth when primitive life existed (Cockell *et al.*, 2000), so UV levels alone may not contraindicate martian life forms.

However, the high UV levels together with the highly oxidizing chemistry of the martian soils is an extremely destructive combination for life forms (Biemann *et al.*, 1977). Very low levels of organic materials were detected in the martian regolith by the Viking landers which suggests that organic material is being destroyed as fast as or faster than it is being produced on Mars. Highly acidic conditions, such as revealed in Meridiani Planum by Opportunity, could host specialized life forms, but it is unclear if prebiotic chemical reactions could occur in such conditions to give rise initially to life forms (Squyres *et al.*, 2004c).

The combination of low temperature, low pressure, lack of liquid water, high radiation and UV levels, and oxidizing soil conditions makes the martian surface a very inhospitable place for life forms similar to those found on Earth. However, the subsurface might provide conditions more conducive to biologic activity (Clark, 1998; Farmer and Des Marais, 1999; Cockell and Barlow, 2002; Diaz and Schulze-Makuch, 2006). Overlying surface materials provide protection from the hazardous radiation environment. Temperatures increase with depth in the martian substrate and liquid water oases could exist where a high geotherm encounters $H_2O$-rich layers. Long-lived hydrothermal activity associated with volcanism or large impact craters could provide habitable oases for extended periods of time (Newsom *et al.*, 2001; Varnes *et al.*, 2003; Abramov and Kring, 2005). It is within these regions where extant martian life may occur.

## 8.2 Viking biology experiments

The Mariners 4, 6, and 7 missions had led scientists to conclude that Mars was a dry, dead world where life would never have had an opportunity to arise. The Mariner 9 discovery of channels formed by flowing water changed this perception – if water had flowed on Mars in the past, life may have arisen. The 1976 Viking missions therefore included landers with experiments to determine if the martian surface might contain life. The landers used their robotic arms to scoop up soil samples which were then deposited into the experiments (Figure 8.1). The three specific biology experiments were the gas exchange (GEx), pyrolytic release (PR), and labeled

Figure 8.1 The two Viking landers were each equipped with a sampling arm which could retrieve soil samples for analysis of possible microbial life. This sequence of four images shows the Viking 2 lander's sampling arm retrieving a soil sample and the small trench from which the sample was removed. (Image PIA00145, NASA/JPL.)

release (LR) (Figure 8.2). In addition, the gas chromatograph–mass spectrometer (GCMS) was used to look for organic material within the regolith.

The GEx experiment tested the martian regolith for release of biologically derived gases by adding water (Klein, 1977). It utilized two modes – a humid mode and a wet mode when a nutrient solution was added. The experiment tested whether the addition of water or water and nutrients could stimulate dormant organisms in the soil to metabolic activity. In the humid non-nutrient mode, the atmospheric pressure was raised to 200 mbar $(2\times10^4\,\mathrm{Pa})$ and saturated with water vapor. The sample was incubated at temperatures between 281 K and 288 K for seven days under dark conditions. In the wet nutrient mode a concentrated solution of organic and inorganic materials was added to the soil sample and again incubated under 200 mbar $(2\times10^4\ \mathrm{Pa})$ pressure conditions and temperatures in the range of 281–288 K. The latter experiment was run three times for periods ranging from 31 to 200 sols. In both cases, gases were released by desorption from the soil and oxygen was generated. However, these gases were likely produced by chemical reactions of the oxidized martian soil with the water and nutrients, not by biologic activity.

The PR experiment (also called the carbon assimilation experiment) approximated the conditions on Mars and incubated the samples under both dark and light conditions (Horowitz *et al.*, 1977; Klein, 1977). Experiments under lighting conditions removed wavelengths shorter than 320 nm. No nutrients were added in either mode of operation, but a trace of water was added in one test. The test was designed to determine if microorganisms in the soil directly synthesize carbon from the atmosphere (CO and/or $CO_2$) and emit gases as waste products. Gases were emitted

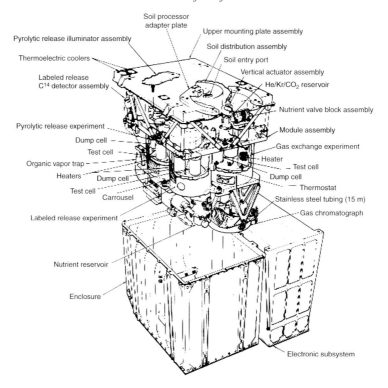

Figure 8.2 This diagram shows the layout of the Viking biology experiment package. The labeled release (LR), pyrolytic release (PR), gas exchange (GEx), and gas chromatograph–mass spectrometer (GCMS) were the experiments which analyzed martian soil samples for evidence of life. (NASA.)

during the test. However, heating of the sample to 913 K, which should have sterilized it, and retesting revealed that approximately the same amount of gases were produced. The thermal stability of these experimental results strongly supports a non-biologic chemical reaction rather than biologic metabolism as the source.

The LR experiment added water containing a dilute solution of simple organic compounds to the soil sample, then incubated the sample between 13 and 90 sols at a temperature of 283 K and pressure of ~60 mbar. Gases were again detected from the incubated soil sample. The temperature of the sample was then raised to 433 K in order to sterilize it and retested. The amount of gas released was lower than the initial amount but gas was still given off. The results of the LR could be consistent either with metabolizing organisms in the soil or by non-biologic chemical reactions (Klein, 1977).

The GCMS experiment analyzed the composition of the martian regolith samples. Its search for organic materials within the samples was negative (Biemann *et al.*,

1977). Exposure of surface materials to solar UV combined with the oxidized soil chemistry likely destroys any organic material in relatively short periods of time.

The results of the Viking lander experiments argue against microorganisms in the martian regolith. Results of all of the experiments are consistent with desorption of atmospheric gases retained by the soil and/or chemical reactions resulting from the oxidized nature of the soil. Most scientists believe the Viking results indicate that life does not exist on the surface of Mars at the present time (Space Studies Board, 1977), consistent with our understanding of life under the hostile conditions described in Section 8.1. However, the landing sites of the two Viking landers were not in regions that would be optimal for life to exist. Therefore much of the recent effort at determining if martian life currently exists focuses on "special regions" where conditions may be more conducive (Klein, 1998; Beaty *et al.*, 2006).

## 8.3 Martian meteorite ALH84001

The Viking biology experiments indicated that extant martian life is unlikely on the planet's surface, but they did not address the possibility of ancient martian life. Interest in the possibility of extinct martian life resurged in 1996 following the announcement that chemical evidence within the ALH84001 martian meteorite was consistent with ancient bacteria (McKay *et al.*, 1996). ALH84001 (Figure 8.3) is the one martian meteorite with an orthopyroxenite composition and a 4.5 Ga crystallization age (Mittlefehldt, 1994). It contains large globules of carbonates (Figure 4.11) (Romanek *et al.*, 1994) which formed ~3.9 Ga ago (Borg *et al.*, 1999). Within

Figure 8.3 ALH84001 is the only martian meteorite currently in our collections which samples the ancient cumulate crust of Mars. Mineralogic evidence within the meteorite led scientists in 1996 to suggest that the meteorite contained evidence of ancient martian bacteria. (NASA/JSC.)

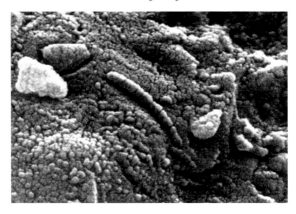

Figure 8.4 Electronic microscopic images of the ALH84001 carbonate revealed a number of small elongated features which McKay *et al.* (1996) interpreted as fossilized nanobacteria. The diameter of this feature is about 1/100 the width of a human hair. (NASA/JSC.)

the carbonates, McKay *et al.* (1996) found polycyclic aromatic hydrocarbons (PAHs), magnetite, and iron sulfides which they argued were formed by martian bacteria. They also noted small features (Figure 8.4) which they suggested were fossilized nanobacteria.

The ALH84001 evidence has been the topic of many heated debates among the scientific community since it was announced. At present, most scientists believe that the PAHs and iron sulfides are either non-biologic in origin or result from terrestrial contamination (Scott, 1999; Zolotov and Shock, 2000). The "nanofossils" have been determined to be either an artifact from the preparation of the sample for scanning electron microscope analysis or a terrestrial weathering effect (Sears and Kral, 1998). The strongest evidence remaining for a biogenic origin of the ALH84001 mineralogies is the magnetite, whose shape is consistent with an origin by bacteria (Thomas-Keprta *et al.*, 2000). However, most scientists believe that the evidence in ALH84001 is too weak to support the assertion that it contains evidence of ancient martian biologic activity.

## 8.4 Atmospheric methane

Ground-based observations (Krasnopolsky *et al.*, 2004) and measurements from MEx's Planetary Fourier Spectrometer (PFS) (Formisano *et al.*, 2004) have detected methane ($CH_4$) in the martian atmosphere with a global average of $\sim 10$ parts per billion (ppb). However, PFS detected regional variations in the concentration of $CH_4$, ranging between 0 and 35 ppb, suggesting localized sources/sinks of the gas (Formisano *et al.*, 2004). Methane undergoes photochemical losses in the martian atmosphere, with a mean loss rate of $2.2 \times 10^5 \, \mathrm{cm^{-2} s^{-1}}$ and a mean lifetime of $\sim 340$ years (Krasnopolsky *et al.*, 2004; Krasnopolsky, 2006), indicating that it must constantly be replenished. The timescale of global mixing of $CH_4$ is $\sim 0.5 \, \mathrm{yr}$, so constant replenishment also is indicated by the regional variations in concentration (Krasnopolsky *et al.*, 2004).

The 1700 ppb of terrestrial atmospheric methane primarily results from biologic activity and methane-loving organisms (methanogens) are plausible candidates for martian microbes if they exist (Max and Clifford, 2000; Jakosky *et al.*, 2003; Varnes *et al.*, 2003). If martian atmospheric $CH_4$ results from subsurface biological activity, this microbial population must be much smaller than on Earth because of the lower $CH_4$ concentrations. Krasnopolsky *et al.* (2004) estimated that the dry biomass produced by biologic methanogenesis on Mars is between $15 \times 10^{-13}$ and $4 \times 10^{-14}$ $\mathrm{kg \, m^{-3}}$, well below the detection limit of $10^{-8} \mathrm{kg \, m^{-3}}$ of Viking's GCMS.

Biologic activity is not the only possible source of martian atmospheric $CH_4$. Dust and chemical changes to the atmosphere produced during cometary impacts can produce $CH_4$ (Kress and McKay, 2004), but the amounts are too low to explain the globally averaged concentrations of 10 ppb (Krasnopolsky, 2006). Volcanic outgassing is a possible source, but $CH_4$ is a minor component of terrestrial volcanic emissions (Ryan *et al.*, 2006) and no active volcanism has been detected on Mars from either visible or thermal IR analysis. Low-temperature alteration of basalt by carbon-rich hydrothermal fluids is a possible abiogenic source of $CH_4$ (Lyons *et al.*, 2005), as is photolysis of $H_2O$ in the presence of CO (Bar-Nun and Dimitrov, 2006). Carbon isotopic analysis is necessary to determine whether the $CH_4$ results from biogenic or abiogenic processes.

## 8.5 Future missions

The Viking experiments indicate that life is not widely distributed across the martian surface. However, there are localized areas that might be more conducive to life forms. In addition, past conditions may have been more hospitable and traces of extinct life might be retained in certain areas of the planet. Thus the question of martian life (extinct or extant) still remains to be answered unconditionally. Future missions are being planned which will help address these issues.

The first post-Viking mission specifically designed to look for evidence of martian life was the Beagle 2 lander (Figure 8.5), which separated from the Mars Express orbiter in December 2003. Beagle 2 carried instruments designed to identify water in the soil, rocks, and atmosphere and traces of life through direct measurement of carbon compounds (Chicarro, 2002). It landed in the Isidis Basin but no signal was returned from the spacecraft after landing.

Figure 8.5 Artist's concept of the Beagle 2 lander after deployment at its Isidis Planitia landing site. Unfortunately no signal was ever received from Beagle 2. (Image ESA20MGBCLC, ESA/Denman Productions.)

Figure 8.6 Phoenix is scheduled to land on Mars in 2008 and will investigate the martian soil and its ability to contain biologic material. This artist's concept shows Phoenix after landing in the northern plains. (NASA/JPL.)

The next mission that will include instruments to look for evidence of life is NASA's Phoenix mission (Figure 8.6), which was launched on 4 August 2007. The objectives of the Phoenix mission are (1) study the history of water in all of its phases, and (2) search for evidence of habitable zones and assess the biological potential of the ice–soil boundary (Smith and the Phoenix Science Team, 2004). To achieve its objectives, Phoenix will land in May 2008 within the 65° to 70°N latitude zone, which has been characterized as a region of high near-surface ice content by GRS. Its instruments will investigate subsurface $H_2O$ and its interaction with the atmosphere as well as characterize the composition of the martian regolith

Figure 8.7 The Mars Science Laboratory is expected to arrive on Mars in 2010 and will contain instruments designed to look for evidence of current or past martian life. (Image PIA04892, NASA/JPL.)

down to 0.5 meter depth. The Thermal and Evolved Gas Analyzer (TEGA) instrument will determine the isotopic ratios of hydrogen, oxygen, carbon, and nitrogen within the soil and provide constraints on the role of biological processes in forming these isotopes.

NASA plans to launch the Mars Science Laboratory (MSL) (Figure 8.7) in the fall of 2009 with arrival in October 2010. MSL is a larger, more robust version of the Mars Exploration Rovers currently operating on the planet. MSL will characterize the biology, geology, geochemistry, and radiation environment of its landing location (Vasavada and the MSL Science Team, 2006). Its specific objectives relative to biology include determining the nature and inventory of organic carbon compounds, making an inventory of the chemical building blocks of life, and identifying features which may be representative of biological processes. To achieve these objectives, it will collect rock and soil samples for distribution among on-board test chambers. These experiments will determine the elemental composition of the samples, including organic compounds such as proteins and amino acids which could be produced by life.

ESA has recently selected the ExoMars mission as its first flagship mission in the Aurora program of Moon and Mars exploration (Figure 8.8). ExoMars will launch in 2013 with arrival at Mars in 2015. It will contain a rover and a small fixed surface station (Vago *et al.*, 2006). The rover will carry the Pasteur science package designed to investigate the geology and exobiological properties of the planet. Proposed instruments include cameras, spectrometers (IR, Mössbauer, and Raman), a ground-penetrating radar, X-ray diffractometer, Mars Organics and Oxidants Detector, GCMS, radiation detector, and a meteorological package. An important part of the rover is the incorporation of a drill which will be able to access the subsurface to a depth of 2 m to explore the possibility of subsurface biologic oases. The surface

Figure 8.8 ESA's ExoMars mission will consist of a rover and a fixed surface station. The rover, shown in this artist's conception, will be able to drill to depths of 2 m to determine if life might exist underground. (ESA/AOES Medialab.)

station will contain the Geophysics/Environment Package (GEP), although the specific instruments have not been selected.

Other missions focused on searching for evidence of martian biology are being discussed. These include additional landers and sample return missions. In addition, many studies are being conducted in remote regions of Earth to characterize the extreme conditions under which terrestrial lifeforms survive and reproduce. These include dry, cold environments such as the Dry Valleys region of Antarctica (Doran *et al.*, 1998; Wentworth *et al.*, 2005; Abyzov *et al.*, 2006), volcanic regions where thermophilic organisms might prosper (Farmer, 1998; Bishop *et al.*, 2004), sulfur-rich volcanic environments (Fernández-Remolar *et al.*, 2005; Knoll *et al.*, 2005), and salt-rich environments (Mancinelli *et al.*, 2004; Reid *et al.*, 2006). These studies help us to better understand the environments under which terrestrial organisms have flourished and provide us with insights into possible astrobiological niches on Mars.

## 8.6 Planetary protection issues

The possibility that Mars may possess extant life forms or at least provide environments where life forms could survive and replicate leads to issues of planetary protection. The basic tenet of planetary protection is that exploration must not jeopardize possible life forms throughout the Solar System. The Committee on Space Research (COSPAR) is an international organization charged with developing and maintaining planetary protection policies (Rummel *et al.*, 2002). Contamination can be either from Earth-based organisms being transported to another world ("forward contamination") or from extraterrestrial organisms being transported to Earth ("backward contamination"). COSPAR defines five categories of planetary protection. Category I includes any type of mission to a body with low probability of life. Examples include Venus and the Moon. Category II are bodies which may provide insights into the origin of life but which provide only a remote chance that contamination by a spacecraft will jeopardize future exploration. Examples are comets and the Jovian planets. Mars exploration is included in Categories III and IV since it is an environment where forward contamination would jeopardize future biological explorations. Category III refers to flyby and orbiter missions while Category IV covers landers and atmospheric probes. Category V refers to a mission from any body which is returning samples to the Earth and is concerned with backward contamination.

Mars orbiters are required to meet the Viking lander pre-sterilization spore levels in the event that the spacecraft crashes into the planet (Rummel and Meyer, 1996). The Viking landers were assembled in a Class 100000 clean room with an average spore burden of $\leq 300$ spores per square meter and a total spore burden of $\leq 3 \times 10^5$ spores per square meter (DeVincenzi *et al.*, 1996). These are the bioburden levels which Class III spacecraft to Mars are required to meet.

Class IV spacecraft include landers, rovers, penetrators, and atmospheric probes. They are subdivided into Classes IVa and IVb, depending on whether the mission contains life-detection experiments (DeVincenzi *et al.*, 1996). Landers without life-detection experiments require pre-sterilization Viking lander bioburden levels, similar to the Class III spacecraft. Category IVb landers carry life-detection experiments and require Viking-level sterilization levels. The Viking landers were heat sterilized by being exposed to a temperature of $\sim 385$ K for 30 hours.

The 2005 COSPAR planetary protection policy defined a Class IVc mission, which are those landers investigating special regions on Mars. A special region is defined as one where terrestrial organisms could propagate or which has a high potential for extant martian life (see COSPAR, 2005). The Mars Special Regions Science Analysis Group (SR-SAG) (Beaty *et al.*, 2006) rated the potential of specific martian environments to exceed the threshold conditions in temperature (253 K) and water activity (0.5) necessary for microbial propagation within $\sim 5$ m of the surface for a time period of 100 years after arrival of the mission. Both normal and crash landings were considered by the team. Their study finds that the only regions that show a significant probability of association with modern liquid water were the gullies and the pasted-on mantle material in the $\sim 30°$–$80°$ latitude zone in both hemispheres. The recent detection by MGS of changes in gully appearance which may indicate recent water flow reinforces the special region nature of these features (Malin *et al.*, 2006) (Figure 8.9).

Regions for which there is a low but non-zero probability of modern liquid water include the low-latitude slope streaks, low-latitude features proposed to be glaciers, and features hypothesized to be deposits of massive subsurface ice. Features that

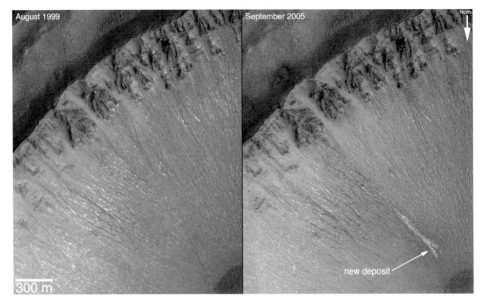

Figure 8.9 Gully formation continues into the present day, as these two images of a crater in the Centauri Montes region demonstrate. A light-colored deposit formed within the 6-year period between August 1999, when the left image was acquired, and September 2005 when the right image was taken. Although the deposit might have been formed by a dust avalanche, water flow is the preferred interpretation. (MOC image MOC2-1619, NASA/JPL/MSSS.)

could have a high probability of association with modern liquid water but which have not been identified as existing at the present time include young volcanic environments, possible hydrothermal systems in recent large impact craters, and modern outflow channels. Overall, the SR-SAG found that while conditions might support the existence of life for short time periods, the conditions on Mars were too hostile to allow such life to replicate which would lead to large-scale biological contamination.

Future missions to Mars include plans for sample return. This raises the small but non-zero chance that martian microorganisms might be brought to Earth which could prove dangerous to terrestrial life (Race, 1996; Trofimov *et al.*, 1996). The 2005 COSPAR Planetary Protection Policy states that samples which will not be subject to life-detection investigations should be sterilized before return to Earth. Unsterilized samples must be placed in a sample container which is then sealed and placed in another containment system whose integrity will be verified before the sample is returned to Earth. This latter containment system must not have contacted Mars, either directly or indirectly, thus breaking the "chain of contact" between Mars and Earth. No uncontained hardware or samples which contacted Mars and which have not been sterilized can be returned to Earth. Once on Earth, all samples

Figure 8.10 The Lunar Receiving Laboratory was built at NASA's Johnson Space Center as a containment facility in which to store and analyze the lunar samples without cross-contamination. Similar laboratory designs are being evaluated for the return of samples from Mars. (NASA/JSC.)

will be sent to receiving facilities designed to contain the material and allocate it for scientific study, similar to the process instituted at NASA Johnson Space Center for receiving, cataloging, and distributing the lunar samples returned by the Apollo missions (Mangus and Larsen, 2004) (Figure 8.10).

# 9

# Looking ahead

Our paradigm of Mars has shifted several times, particularly during the past 40 years of spacecraft observations. Lowell's view of a world criss-crossed by canals built by a dying race of Martians was supplanted in the 1960s by Mariner views of a geologically dead world with an atmosphere too thin to support liquid water. Mariner 9's discovery of channels and volcanoes shifted the view to one of a planet where water existed in the past and led to renewed interest in the possibility that life might exist on our neighboring world. The Viking lander investigations quashed hopes of finding the martian soil teeming with microbial life, but refocused the interest in biological activity (past or present) to identifying localized oases. Recognition that we have samples of the planet's crust in the form of martian meteorites has greatly influenced our understanding of the planet's bulk chemistry and thermal evolution. The latest shift in our mental picture of Mars has revealed that water, either liquid or ice, has played a larger role throughout the planet's history than previously recognized. Our post-Viking view was that Mars had a thicker atmosphere and warmer, wetting surface conditions early in its history, but the post-Noachian period has been characterized by the cold, dry climate that we see today. Insights gained from MGS, Odyssey, MER, and MEx combined with increasing computational capabilities have revealed that short-lived climate excursions have occurred up through recent times, driven by changes in orbital parameters and the planet's obliquity. Equatorial cold-based glaciers and mid-latitude gullies probably formed by seeping groundwater reveal that water is actively altering the face of Mars even today.

We are entering a new era of Mars exploration. The mineralogic information provided by Spirit and Opportunity combined with orbital investigations from TES, THEMIS, and OMEGA have revealed that the planet has distinct mineralogic periods which generally coincide with the well-established geologic periods. These results further illuminate the geologic and climatic changes that the planet has experienced throughout its 4.5 Ga of existence. The increasingly higher resolutions

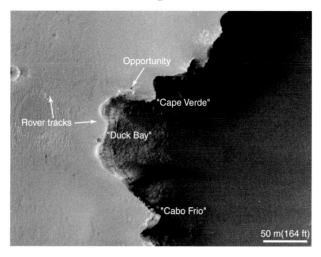

Figure 9.1 The Opportunity rover had just arrived at the rim of Victoria Crater when MRO's HiRISE camera snapped its picture from orbit. This is the first time that a surface mission has been clearly seen in images taken by a Mars orbiter. (PIA08817, NASA/JPL/University of Arizona.)

achieved by spectrometers, such as CRISM on MRO, will permit even more detailed investigations in the upcoming years and allow us to begin to understand the small-scale geology of the planet.

Higher resolutions are not limited just to the spectrometers. MOC, THEMIS, and HRSC have taken us from the kilometer-scale views of earlier orbiters to meter-scale perspectives where localized geological processes begin to reveal themselves. These improved views have helped us narrow the possible range of processes operating in different regions of the planet. The arrival of MRO with the HiRISE camera is taking us to the next step of seeing surface features down to the centimeter scale. For the first time we can actually see our surface landers and rovers from orbit (Figure 9.1), which exemplifies the new perspective we can anticipate of the martian surface.

Topographic information from MOLA and the stereo views of HRSC have provided us with three-dimensional views of martian surface features for the first time. Determining depths, heights, slopes, and roughness helps constrain the types of processes that formed and modified the observed surface features. We can now also peer below the surface through use of MEx's MARSIS and MRO's SHARAD ground-penetrating radars and determine how subsurface structure and composition, particularly the distribution of subsurface ice and liquid, influence the features seen on the martian surface.

Long-term monitoring of the martian atmosphere by TES, PFS, and SPICAM combined with surface monitoring from landers and rovers is helping us to understand the structure of the atmosphere and the processes driving its circulation. Features that had previously been proposed based on general circulation models have now been confirmed by these instruments. Advances in computational capabilities and improved modeling of the terrestrial atmosphere have enhanced the modeling of the martian atmosphere and mesoscale models are providing us with our first detailed understanding of weather and climate at specific locations on the surface. New insights of weather and atmospheric structure can be anticipated as data from MRO's Mars Climate Imager (MARCI) and Mars Climate Sounder (MCS) are analyzed.

Upcoming missions will focus on addressing specific questions about Mars. Phoenix will provide the first detailed analysis of surface ice. Its Microscopy, Electrochemistry, and Conductivity Analyzer (MECA) will determine the mineralogy, pH, water/ice content, and soil/ice conductivity at the landing site. Complementing MECA is the Thermal and Evolved Gas Analyzer (TEGA) experiment which will characterize the chemistry of ice and soil samples and look for evidence of organic molecules which might have been produced by biological activity. The Surface Stereo Imager (SSI) and Robotic Arm Camera (RAC) will determine the geological context of the samples and surroundings, and the Meterological Station (MET) will provide daily weather reports during the lander's operational lifetime.

The Mars Science Laboratory (MSL) is a larger and heavier version of the Mars Exploration Rovers, with enhanced capabilities for on-board analysis of surface materials. It will include a laser to vaporize thin surface layers of rock and allow chemical analysis of the rock's interior. On-board test chambers will analyze the chemistry of soil and crushed rock samples, including detection of organic molecules indicative of life. A neutron spectrometer will be used to locate water and a meterological package will monitor the atmospheric conditions and daily weather as the rover moves across the planet's surface.

Two candidates for the 2011 Mars Scout mission were announced in January 2007. The two candidates, Mars Atmosphere and Volatile Evolution (MAVEN) and The Great Escape (TGE), would study the evolution of Mars' upper atmosphere through analysis of the structure and dynamics operating in this region. A major emphasis of both missions will be to understand atmospheric loss through interactions of the upper atmosphere with the solar wind. NASA will select one of these proposals in early 2008 as the final design for the 2011 Mars Scout mission.

ESA's ExoMars will be the first Mars mission with drilling capabilities. Although primarily advertised as an astrobiology mission (Section 8.5), ExoMars will also characterize the geochemical and geophysical characteristics of its landing area. The Pasteur experiment package will investigate the interior composition of rocks encountered by the rover and can drill up to 2 meters in the soil to collect samples which will be analyzed by on-board experiments. The lander's Geophysics/Environmental

Package will supplement the data acquired by other landers and rovers about the weather and surface material properties.

These upcoming missions will address some of the current questions about Mars' upper atmosphere, surface geochemical and geophysical characteristics, and the planet's potential for harboring extinct or extant life. However, answers to many other questions will have to await missions yet to be defined. The interior structure of Mars is one of the biggest uncertainties and will require emplacement of a seismic network across the planet. Experiments to measure heat flow at various locations across the planet will constrain the present-day geothermal flux and provide insights into the stability of near-surface ice and the thermal evolution of the planet. Analysis of martian meteorites has provided our first glimpse of the planet's bulk chemistry, but they largely represent young basaltic flows. Analysis of a wider variety of rock samples from across the martian surface will better define the geochemical evolution of the planet. Age-dating of surface samples is needed to constrain the martian crater chronology and determine its similarities and differences to that derived for the Earth–Moon system. Firm establishment of the martian chronology will further illuminate the geologic, geochemical, thermal, and climatic histories of the planet.

A common question about the future exploration of Mars is the relative role of robots versus humans. Robotic missions are cheaper than human missions and can operate for longer periods, typically until the mission reaches the end of its operating lifetime. Advances in technology are producing rover and lander missions capable of feats that were only dreamed of during the Viking era. However, the Apollo missions taught us that trained astronauts can easily outperform a robotic mission. Spirit took five months to travel approximately 3 km from its landing site on the Gusev plains to Columbia Hills. Much of that time was spent driving and if the rover passed an unusual rock during its periods of autonomous driving, the rock went unnoticed. Humans not only could travel greater distances in shorter periods of time but also can climb steeper slopes and stop to investigate serendipitous discoveries along the way. Human missions also are typically of more interest to those remaining behind on Earth, notwithstanding the incredible Internet traffic associated with the MPF and MER missions. However, humans require expensive life-support (air, food, water, shelter) and the extent of such a mission will be shorter because of the need to return to Earth. Sending humans to Mars, as proposed in NASA's Vision for Space Exploration, will take substantial resources and commitment to achieve. International participation, and sharing of the costs, will likely be necessary to take this step. Robotic missions cannot be sacrificed to pay the way for humans – a balanced program of robotic and human exploration will provide the best return on the monetary investment.

I am often asked why we should spend the money exploring Mars when there are so many problems facing us here on Earth. My response is that the technologies

Figure 9.2 The Earth's clouds and water-covered surface contrast with the dry conditions and thin atmosphere of Mars. (From *Viking Orbiter Views of Mars*, NASA SP-441, courtesy Lunar and Planetary Institute.)

developed for studying Mars and the knowledge that we gain from our neighboring world benefit all of us on Earth. Mars exploration excites people, particularly our youth. Such excitement in the early days of the space program led those of us currently involved in Mars exploration to pursue scientific and engineering careers. Our current discoveries are inspiring a new generation of scientists and engineers whose creativity and inventions will enable breakthroughs to resolve many of the problems facing humanity. The conditions on present-day Mars contrast dramatically with the warm, water-rich Earth, enhancing our awareness of our own fragile environment (Figure 9.2). Our growing realization that early Mars was more Earth-like with liquid water, a thicker atmosphere, and warmer surface conditions leads me and many of my colleagues to investigate what happened to produce the cold, dry planet we see at present. Such knowledge will help us be better caretakers of the blue orb that we call home.

# References

Abe, Y. (1997). Thermal and chemical evolution of the terrestrial magma ocean. *Physics of the Earth and Planetary Interiors*, **100**, 27–39.

Abramov, O. and Kring, D.A. (2005). Impact-induced hydrothermal activity on early Mars. *Journal of Geophysical Research*, **110**, E12S09, doi: 10.1029/2005JE002453.

Abyzov, S.S., Duxbury, N.S., Bobin, N.E., *et al.* (2006). Super-long anabiosis of ancient microorganisms in ice and terrestrial models for development of methods to search for life on Mars, Europa, and other planetary bodies. *Advances in Space Research*, **38**, 1191–1197.

Acuña, M.H., Connerney, J.E.P., Wasilewski, P., *et al.* (1998). Magnetic field and plasma observations at Mars: initial results of the Mars Global Surveyor Mission. *Science*, **279**, 1676–1680.

Acuña, M.H., Connerney, J.E.P., Ness, N.F., *et al.* (1999). Global distribution of crustal magnetization discovered by the Mars Global Surveyor MAG/ER Experiment. *Science*, **284**, 790–793.

Acuña, M.H., Connerney, J.E.P., Wasilewski, P., *et al.* (2001). Magnetic field of Mars: summary of results from the aerobraking and mapping orbits. *Journal of Geophysical Research*, **106**, 23403–23417.

Agee, C.B. and Draper, D.S. (2004). Experimental constraints on the origin of martian meteorites and the composition of the martian mantle. *Earth and Planetary Science Letters*, **224**, 415–429.

Agnor, C.B., Canup, R.M., and Levison, H. (1999). On the character and consequences of large impacts in the late stage of terrestrial planet formation. *Icarus*, **142**, 219–237.

Albee, A.L., Arvidson, R.E., and Palluconi, F.D. (1992). Mars Observer Mission. *Journal of Geophysical Research*, **97**, 7665–7680.

Albee, A.L., Arvidson, R.E., Palluconi, F., and Thorpe, T. (2001). Overview of the Mars Global Surveyor mission. *Journal of Geophysical Research*, **106**, 23291–23316.

Alves, E.I. and Baptista, A.R. (2004). Rock magnetic fields shield the surface of Mars from harmful radiation. In *Lunar and Planetary Science XXXV*, Abstract #1540. Houston, TX: Lunar and Planetary Institute.

Anderson, D.L., Miller, W.F., Latham, G.V., *et al.* (1977). Seismology on Mars. *Journal of Geophysical Research*, **82**, 4524–4546.

Anderson, R.C., Dohm, J.M., Golombek, M.P., *et al.* (2001). Primary centers and secondary concentrations of tectonic activity through time in the western hemisphere of Mars. *Journal of Geophysical Research*, **106**, 20563–20586.

Anderson, R.C., Dohm, J.M., Haldemann, A.F.C., Pounders, E., and Golombek, M.P. (2006). Tectonic evolution of Mars. In *Lunar and Planetary Science XXXVII*, Abstract #1883. Houston, TX: Lunar and Planetary Institute.

Anguita, F., Anguita, J., Castilla, G., *et al.* (1998). Arabia Terra, Mars: tectonic and paleoclimatic evolution of a remarkable sector of martian lithosphere. *Earth, Moon and Planets*, **77**, 55–72.

Anguita, F., Babin, R., Benito, G., *et al.* (2000). Chasma Australe, Mars: structural framework for a catastrophic outflow origin. *Icarus*, **144**, 302–312.

Arfstrom, J. and Hartmann, W.K. (2005). Martian flow features, moraine-like ridges, and gullies: terrestrial analogs and interrelationships. *Icarus*, **174**, 321–335.

Arkani-Hamed, J. (2001). Paleomagnetic pole positions and pole reversals of Mars. *Geophysical Research Letters*, **28**, 3409–3412.

Arvidson, R.E., Anderson, R.C., Bartlett, P., *et al.* (2004a). Localization and physical properties experiments conducted by Spirit at Gusev Crater. *Science*, **305**, 821–824.

Arvidson, R.E., Anderson, R.C., Bartlett, P. *et al.* (2004b). Localization and physical properties experiments conducted by Opportunity at Meridiani Planum. *Science*, **306**, 1730–1733.

Badhwar, G.D. (2004). Martian Radiation Environment Experiment (MARIE). *Space Science Reviews*, **110**, 131–142.

Baker, V.R. (1978). The Spokane flood controversy and the martian outflow channels. *Science*, **202**, 1249–1256.

Baker, V.R. (1982). *The Channels of Mars*. Austin, TX: University of Texas Press.

Baker, V.R. (2001). Water and the martian landscape. *Nature*, **412**, 228–236.

Baker, V.R., Strom, R.G., Gulick, V.C., *et al.* (1991). Ancient oceans, ice sheets, and the hydrological cycle on Mars. *Nature*, **352**, 589–594.

Balsiger, H., Altwegg, K., and Geiss, J. (1995). D/H and $^{18}O/^{16}O$ ratio in the hydronium ion and in neutral water from in situ ion measurements in Comet P/Halley. *Journal of Geophysical Research*, **100**, 5827–5834.

Bandfield, J.L. (2002). Global mineral distributions on Mars. *Journal of Geophysical Research*, **107**, 5042, doi: 10.1029/2001JE001510.

Bandfield, J.L., Hamilton, V.E., and Christensen, P.R. (2000). A global view of martian volcanic compositions. *Science*, **287**, 1626–1630.

Bandfield, J.L., Hamilton, V.E., Christensen, P.R., and McSween, H.Y. (2004). Identification of quartzofeldspathic materials on Mars. *Journal of Geophysical Research*, **109**, E10009, doi: 10.1029/2004JE002290.

Banerdt, W.B., Golombek, M.P., and Tanaka, K.L. (1992). Stress and tectonics on Mars. In *Mars*, ed. H.H. Kieffer, B.M. Jakosky, C.W. Snyder, and M.S. Matthews. Tucson, AZ: University of Arizona Press, pp. 249–297.

Banfield, D., Conrath, B., Pearl, J.C., Smith, M.D., and Christensen, P. (2000). Thermal tides and stationary waves on Mars as revealed by Mars Global Surveyor Thermal Emission Spectrometer. *Journal of Geophysical Research*, **105**, 9521–9537.

Banfield, D., Conrath, B.J., Smith, M.D., Christensen, P.R., and Wilson, R.J. (2003). Forced waves in the martian atmosphere from MGS TES nadir data. *Icarus*, **161**, 319–345.

Banfield, D., Conrath, B.J., Gierasch, P.J., Wilson, R.J., and Smith, M.D. (2004). Traveling waves in the martian atmosphere from MGS TES nadir data. *Icarus*, **170**, 365–403.

Barlow, N.G. (1988). Crater size–frequency distributions and a revised martian relative chronology. *Icarus*, **75**, 285–305.

Barlow, N.G. (1990). Constraints on early events in Martian history as derived from the cratering record. *Journal of Geophysical Research*, **95**, 14191–14201.

Barlow, N.G. (1997). Identification of possible source craters for Martian meteorite ALH84001. In *Instruments, Methods, and Missions for the Investigation of Extraterrestrial Microorganisms*, ed. R.B. Hoover. Bellingham, WA: SPIE Proceedings vol. 3111, pp.26–35.

Barlow, N.G. (2004). Martian subsurface volatile concentrations as a function of time: clues from layered ejecta craters. *Geophysical Research Letters*, **31**, L05703, doi: 10.1029/2003GL019075.

Barlow, N.G. (2005). A review of martian impact crater ejecta structures and their implications for target properties. In *Large Meteorite Impacts III*, ed. T. Kenkmann, F. Hörz, and A. Deutsch. Boulder, CO: Geological Society of America Special Paper 384, pp.433–442.

Barlow, N.G. (2006). Impact craters in the northern hemisphere of Mars: layered ejecta and central pit characteristics. *Meteoritics and Planetary Science*, **41**, 1425–1436.

Barlow, N.G. and Perez, C.B. (2003). Martian impact crater ejecta morphologies as indicators of the distribution of subsurface volatiles. *Journal of Geophysical Research*, **108**, 5085, doi: 10.1029/2002JE002036.

Barlow, N.G., Boyce, J.M., Costard, F.M., *et al.* (2000). Standardizing the nomenclature of martian impact crater ejecta morphologies. *Journal of Geophysical Research*, **105**, 26733–26738.

Barnes, J.R., Pollack, J.B., Haberle, R.M., *et al.* (1993). Mars atmospheric dynamics as simulated by the NASA/Ames general circulation model. 2. Transient baroclinic eddies. *Journal of Geophysical Research*, **98**, 3125–3148.

Barnes, J.R., Haberle, R.M., Pollack, J.B., Lee, H., and Schaeffer, J. (1996). Mars atmospheric dynamics as simulated by the NASA/Ames general circulation model. 3. Winter quasistationary eddies. *Journal of Geophysical Research*, **101**, 12753–12776.

Barnouin-Jha, O.S., Schultz, P.H., and Lever, J.H. (1999a). Investigating the interactions between an atmosphere and an ejecta curtain. 1. Wind tunnel tests. *Journal of Geophysical Research*, **104**, 27105–27115.

Barnouin-Jha, O.S., Schultz, P.H., and Lever, J.H. (1999b). Investigating the interactions between an atmosphere and an ejecta curtain. 2. Numerical experiments. *Journal of Geophysical Research*, **104**, 27117–27131.

Bar-Nun, A. and Dimitrov, V. (2006). Methane on Mars: a product of $H_2O$ photolysis in the presence of CO. *Icarus*, **181**, 320–322.

Basu, S., Richardson, M.I., and Wilson, R.J. (2004). Simulation of the martian dust cycle with the DFDL Mars GCM. *Journal of Geophysical Research*, **109**, E11006, doi: 10.1029/2004JE002243.

Batson, R.M., Edwards, K., and Duxbury, T.C. (1992). Geodesy and cartography of the Martian satellites. In *Mars*, ed. H.H. Kieffer, B.M. Jakosky, C.W. Snyder, and M.S. Matthews. Tucson, AZ: University of Arizona Press, pp.1249–1256.

Beaty, D., Buxbaum, K., Meyer, M., *et al.* (2006). Findings of the Mars Special Regions Science Analysis Group. *Astrobiology*, **6**, 677–732.

Bell, J.F., Pollack, J.B., Geballe, T.R., Cruikshank, D.P., and Freedman, R. (1994). Spectroscopy of Mars from 2.04 to 2.44 μm during the 1993 opposition: absolute calibration and atmospheric vs. mineralogic origin of narrow absorption features. *Icarus*, **111**, 106–123.

Bell, J.F., Wolff, M.J., James, P.B., *et al.* (1997). Mars surface mineralogy from Hubble Space Telescope imaging during 1994–1995: observations, calibration, and initial results. *Journal of Geophysical Research*, **102**, 9109–9123.

Bell, J.F., McSween, H.Y., Crisp, J.A., *et al.* (2000). Mineralogic and compositional properties of martian soil and dust: results from Mars Pathfinder. *Journal of Geophysical Research*, **105**, 1721–1755.

Belleguic, V., Lognonné, P., and Wieczorek, M. (2005). Constraints on the martian lithosphere from gravity and topography data. *Journal of Geophysical Research*, **110**, E11005, doi: 10.1029/2005JE002437.

Benson, J.L. and James, P.B. (2005). Yearly comparisons of the martian polar caps: 1999–2003 Mars Orbiter Camera observations. *Icarus*, **174**, 513–523.

Benz, W. and Cameron, A.G.W. (1990). Terrestrial effects of the giant impact. In *Origin of the Earth*, ed. H.E. Newsom and J.H. Jones. Oxford, UK: Oxford University Press, pp.61–67.

Benz, W., Cameron, A.G.W., and Melosh, H.J. (1989). The origin of the Moon and the single-impact hypothesis III. *Icarus*, **81**, 113–131.

Berman, D.C. and Hartmann, W.K. (2002). Recent fluvial, volcanic, and tectonic activity on the Cerberus Plains of Mars. *Icarus*, **159**, 1–17.

Berman, D.C., Crown, D.A., and Hartmann, W.K. (2005). Tyrrhena Patera, Mars: insights into volcanic and erosional history from high-resolution images and impact crater populations. *American Geophysical Union Fall Meeting*, Abstract #P23A-0177.

Bertelsen, P., Goetz, W., Madsen, M.B., *et al.* (2004). Magnetic properties experiments on the Mars Exploration Rover Spirit at Gusev Crater. *Science*, **305**, 827–829.

Bertka, C.M. and Fei, Y.J. (1997). Mineralogy of the martian interior up to core–mantle boundary pressures. *Journal of Geophysical Research*, **102**, 5251–5264.

Beyer, R.A. and McEwen, A.S. (2005). Layering stratigraphy of eastern Coprates and northern Capri Chasmata, Mars. *Icarus*, **179**, 1–23.

Bhattacharya, J.P., Payenberg, T.H.D., Lang, S.C., and Bourke, M. (2005). Dynamic river channels suggest a long-lived Noachian crater lake on Mars. *Geophysical Research Letters*, **32**, L10201, doi: 10.1029/2005GL022747.

Bibring, J.-P., Langevin, Y., Poulet, F., *et al.* (2004). Perennial water ice identified in the south polar cap of Mars. *Nature*, **428**, 627–630.

Bibring, J.-P., Langevin, Y., Gendrin, A., *et al.* (2005). Mars surface diversity as revealed by the OMEGA/Mars Express observations. *Science*, **307**, 1576–1581.

Bibring, J.-P., Langevin, Y., Mustard, J.F., *et al.* (2006). Global mineralogical and aqueous Mars history derived from OMEGA/Mars Express data. *Science*, **312**, 400–404.

Biemann, K., Oro, J., Toulmin, P., *et al.* (1977). The search for organic substances and inorganic volatile compounds in the surface of Mars. *Journal of Geophysical Research*, **82**, 4641–4658.

Bierhaus, E.B., Chapman, C.R., and Merline, W.J. (2005). Secondary craters on Europa and implications for cratered surfaces. *Nature*, **437**, 1125–1127.

Birck, J.L. and Allègre, C.J. (1978). Chronology and chemical history of the parent body of basaltic achondrites studied by the $^{87}$Rb–$^{87}$Sr method. *Earth and Planetary Science Letters*, **39**, 37–51.

Bishop, J.L., Murad, E., Lane, M.D., and Mancinelli, R.L. (2004). Multiple techniques for mineral identification on Mars: a study of hydrothermal rocks as potential analogs for astrobiology sites on Mars. *Icarus*, **169**, 311–323.

Blamont, J.E. and Chassefière, E. (1993). First detection of ozone in the middle atmosphere of Mars from solar occultation measurements. *Icarus*, **104**, 324–336.

Blichert-Toft, J., Gleason, J.D., Télouk, P., and Albarède, F. (1999). The Lu–Hf isotope geochemistry of shergottites and the evolution of the martian mantle–crust system. *Earth and Planetary Science Letters*, **173**, 25–39.

Bockelée-Moravan, D., Gautier, D., Lis, D.C., *et al.* (1998). Deuterated water in Comet C/1996 B2 (Hyakutake) and its implications for the origin of comets. *Icarus*, **133**, 147–162.

Bogard, D.D. and Johnson, P. (1983). Martian gases in an Antarctic meteorite? *Science*, **221**, 651–654.

Borg, L.E. and Draper, D.S. (2003). A petrogenetic model for the origin and compositional variation of the martian basaltic meteorites. *Meteoritics and Planetary Science*, **38**, 1713–1731.

Borg, L.E., Nyquist, L.E., Taylor, L.A., Wiesmann, H., and Shih, C.-Y. (1997). Constraints on martian differentiation process from Rb–Sr and Sm–Nd isotopic analyses of the basaltic shergottite QUE94201. *Geochimica et Cosmochimica Acta*, **61**, 4915–4931.

Borg, L.E., Connelly, J.N., Nyquist, L.E. *et al.* (1999). The age of the carbonates in martian meteorite ALH84001. *Science*, **268**, 90–94.

Boss, A.P. (1990). 3D solar nebula models: implications for Earth origin. In *Origin of the Earth*, ed. H.E. Newsom and J.H. Jones. Oxford, UK: Oxford University Press, pp.3–15.

Bottke, W.F., Love, S.G., Tytell, D., and Glotch, T. (2000). Interpreting the elliptical crater populations on Mars, Venus, and the Moon. *Icarus*, **145**, 108–121.

Bottke, W.F., Morbidelli, A., Jedicke, R., *et al.* (2002). Debiased orbital and absolute magnitude distribution of the Near-Earth Objects. *Icarus*, **156**, 399–433.

Bottke, W.F., Durda, D.D., Nesvorný, D., *et al.* (2005). The fossilized size distribution of the main asteroid belt. *Icarus*, **175**, 111–140.

Boynton, W.V., Feldman, W.C., Squyres, S.W., *et al.* (2002). Distribution of hydrogen in the near surface of Mars: evidence for subsurface ice deposits. *Science*, **297**, 81–85.

Boynton, W.V., Feldman, W.C., Mitrofanov, I.G., *et al.* (2004). The Mars Odyssey Gamma-Ray Spectrometer instrument suite. *Space Science Reviews*, **110**, 37–83.

Bradley, B.A., Sakimoto, S.E.H., Frey, H., and Zimbelman, J.R. (2002). Medusae Fossae Formation: new perspectives from Mars Global Surveyor. *Journal of Geophysical Research*, **107**, 5058, doi: 10.1029/2001JE001537.

Brain, D.A. and Jakosky, B.M. (1998). Atmospheric loss since the onset of the martian geologic record: combined role of impact erosion and sputtering. *Journal of Geophysical Research*, **103**, 22689–22694.

Brandon, A.D., Walker, R.J., Morgan, J.W., and Goles, G.G. (2000). Re–Os isotopic evidence for early differentiation of the martian mantle. *Geochimica et Cosmochimica Acta*, **64**, 4083–4095.

Breuer, D. and Spohn, T. (2003). Early plate tectonics versus single-plate tectonics on Mars: evidence from magnetic field history and crust evolution. *Journal of Geophysical Research*, **108**, E75072, doi: 10.1029/2002JE001999.

Breuer, D., Spohn, T., and Wüllner, U. (1993). Mantle differentiation and the crustal dichotomy of Mars. *Planetary and Space Science*, **41**, 269–283.

Breuer, D., Yuen, D.A., and Spohn, T. (1997). Phase transitions in the martian mantle: implications for partially layered convection. *Earth and Planetary Science Letters*, **148**, 457–469.

Bridges, J.C. and Grady, M.M. (2000). Evaporite mineral assemblages in the nakhlite (martian) meteorites. *Earth and Planetary Science Letters*, **176**, 267–279.

Bridges, N.T., Phoreman, J., White, B.R., *et al.* (2005). Trajectories and energy transfer of saltating particles onto rock surfaces: application to abrasion and ventifact formation on Earth and Mars. *Journal of Geophysical Research*, **110**, E12004, doi: 10.1029/2004JE002388.

Buczkowski, D.L. and McGill, G.E. (2002). Topography within circular grabens: implications for polygon origin, Utopia Planitia, Mars. *Geophysical Research Letters*, **29**, 1155, doi: 10.1029/2001GL014100.

Burns, J.A. (1977). Orbital evolution. In *Planetary Satellites*, ed. J. A Burns. Tucson, AZ: University of Arizona Press, pp.113–156.

Burr, D.M., Grier, J.A., McEwen, A.S., and Keszthelyi, L.P. (2002). Repeated aqueous flooding from the Cerberus Fossae: evidence for very recently extant, deep groundwater on Mars. *Icarus*, **159**, 53–73.

Byrne, S. and Ingersoll, A.P. (2003). A sublimation model for martian south polar ice features. *Science*, **299**, 1051–1053.

Byrne, S. and Murray, B.C. (2002). North polar stratigraphy and the paleo-erg of Mars. *Journal of Geophysical Research*, **107**, 5044, doi: 10.1029/2001JE001615.

Cabrol, N.A and Grin, E.A. (1999). Distribution, classification, and ages of martian impact crater lakes. *Icarus*, **142**, 160–172.

Cabrol, N.A. and Grin, E.A. (2001). The evolution of lacustrine environments on early Mars: is Mars only hydrologically dormant? *Icarus*, **149**, 291–328.

Cahoy, K.L., Hinson, D.P., and Tyler, G.L. (2006). Radio science measurements of atmospheric refractivity with Mars Global Surveyor. *Journal of Geophysical Research*, **111**, E05003, doi: 10.1029/2005JE002634.

Cameron, A.G.W. (1997). The origin of the Moon and the single impact hypothesis V. *Icarus*, **126**, 126–137.

Cameron, A.G.W. (2000). Higher-resolution simulations of the giant impact. In *Origin of the Earth and Moon*, ed. R.M. Canup and K. Righter. Tucson, AZ: University of Arizona Press, pp.133–143.

Cantor, B.A., James, P.B., Caplinger, M., and Wolff, M.J. (2001). Martian dust storms: 1999 Mars Orbiter Camera observations. *Journal of Geophysical Research*, **106**, 23653–23687.

Canup, R.M. and Agnor, C.B. (2000). Accretion of the terrestrial planets and the Earth–Moon system. In *Origin of the Earth and Moon*, ed. R.M. Canup and K. Righter. Tucson, AZ: University of Arizona Press, pp.113–129.

Carr, M.H. (1981). *The Surface of Mars*. New Haven, CT: Yale University Press.

Carr, M.H. (1995). The martian drainage system and the origin of valley networks and fretted channels. *Journal of Geophysical Research*, **100**, 7479–7507.

Carr, M.H. (1996). *Water on Mars*. New York: Oxford University Press.

Carr, M.H. and Wänke, H. (1992). Earth and Mars: water inventories as clues to accretional histories. *Icarus*, **98**, 61–71.

Carr, M.H., Crumpler, L.S., Cutts, J.A., *et al.* (1977). Martian impact craters and emplacement of ejecta by surface flow. *Journal of Geophysical Research*, **82**, 4055–4065.

Carruthers, M.W. and McGill, G.E. (1998). Evidence for igneous activity and implications for the origin of a fretted channel in southern Ismenius Lacus, Mars. *Journal of Geophysical Research*, **103**, 31433–31443.

Cassen, P. (1994). Utilitarian models of the solar nebula. *Icarus*, **112**, 405–429.

Chambers, J.E. (2001). Making more terrestrial planets. *Icarus*, **152**, 205–224.

Chambers, J.E. (2004). Planetary accretion in the inner solar system. *Earth and Planetary Science Letters*, **223**, 241–252.

Chambers, J.E. and Wetherill, G.W. (1998). Making the terrestrial planets: N-body integrations of planetary embryos in three dimensions. *Icarus*, **136**, 304–327.

Chan, M.A., Beitler, B., Parry, W.T., Ormö, J., and Komatsu, G. (2004). A possible terrestrial analogue for haematite concretions on Mars. *Nature*, **429**, 731–734.

Chen, J.H. and Wasserburg, G.J. (1986). Formation ages and evolution of Shergotty and its parent planet from U–Th–Pb systematics. *Geochimica et Cosmochimica Acta*, **50**, 955–967.

Chicarro, A. (2002). Mars Express mission and astrobiology. *Solar System Research*, **36**, 487–491.

Christensen, P.R. (1986). The spatial distribution of rocks on Mars. *Icarus*, **68**, 217–238.

Christensen, P.R. (2003). Formation of recent martian gullies through melting of extensive water-rich snow deposits. *Nature*, **422**, 45–47.

Christensen, P.R. and Moore, H.J. (1992). The martian surface layer. In *Mars*, ed. H.H. Kieffer, B.M. Jakosky, C.W. Snyder, and M.S. Matthews. Tucson, AZ: University of Arizona Press, pp. 686–729.

Christensen, P.R., Bandfield, J.L., Smith, M.D., Hamilton, V.E., and Clark, R.N. (2000a). Identification of a basaltic component on the martian surface from Thermal Emission Spectrometer data. *Journal of Geophysical Research*, **105**, 9609–9621.

Christensen, P.R., Bandfield, J.L., Hamilton, V.E., *et al.* (2000b). A thermal emission spectral library of rock-forming minerals. *Journal of Geophysical Research*, **105**, 9735–9739.

Christensen, P.R., Bandfield, J.L., Hamilton, V.E., *et al.* (2001a). Mars Global Surveyor Thermal Emission Spectrometer experiment: investigation description and surface science results. *Journal of Geophysical Research*, **106**, 23823–23871.

Christensen, P.R., Morris, R.V., Lane, M.D., Bandfield, J.L., and Malin, M.C. (2001b). Global mapping of martian hematite mineral deposits: remnants of water-driven processes on early Mars. *Journal of Geophysical Research*, **106**, 23873–23885.

Christensen, P.R., Jakosky, B.M., Kieffer, H.H., *et al.* (2004a). The Thermal Emission Imaging System (THEMIS) for the Mars 2001 Odyssey Mission. *Space Science Reviews*, **110**, 85–130.

Christensen, P.R., Ruff, S.W., Fergason, R.L., *et al.* (2004b). Initial results from the Mini-TES experiment in Gusev Crater from the Spirit rover. *Science*, **305**, 837–842.

Christensen, P.R., McSween, H.Y., Bandfield, J.L., *et al.* (2005). Evidence for magmatic evolution and diversity on Mars from infrared observations. *Nature*, **436**, 504–509.

Clark, B.C. (1998). Surviving the limits to life at the surface of Mars. *Journal of Geophysical Research*, **103**, 28545–28555.

Clark, B.C., Baird, A.K., Rose, H.J., *et al.* (1977). The Viking X-ray Fluorescence experiment: analytical methods and early results. *Journal of Geophysical Research*, **82**, 4577–4594.

Clark, B.C., Baird, A.K., Weldon, R.J., *et al.* (1982). Chemical composition of martian fines. *Journal of Geophysical Research*, **87**, 10059–10067.

Clark, R.N. and Roush, T.L. (1984). Reflectance spectroscopy: quantitative analysis techniques for remote sensing applications. *Journal of Geophysical Research*, **89**, 6329–6340.

Clayton, R.N. (1993). Oxygen isotopes in meteorites. *Annual Reviews of Earth and Planetary Science*, **21**, 115–149.

Clayton, R.N. and Mayeda, T.K. (1996). Oxygen isotope studies of achondrites. *Geochimica et Cosmochimica Acta*, **60**, 1999–2017.

Cleghorn, T.F., Saganti, P.B., Zeitlin, C., and Cucinotta, F.A. (2004). Charged particle dose measurements by the Odyssey/MARIE instrument in Mars orbit and model calculations. In *Lunar and Planetary Science XXXV*, Abstract #1321. Houston, TX: Lunar and Planetary Institute.

Clifford, S.M. (1987). Polar basal melting on Mars. *Journal of Geophysical Research*, **92**, 9135–9152.

Clifford, S.M. (1993). A model for the hydrologic and climatic behavior of water on Mars. *Journal of Geophysical Research*, **98**, 10973–11016.

Clifford, S.M. and Hillel, D. (1983). The stability of ground ice in the equatorial region of Mars. *Journal of Geophysical Research*, **88**, 2456–2474.

Clifford, S.M. and Parker, T.J. (2001). The evolution of the martian hydrosphere: implications for the fate of a primordial ocean and the current state of the northern plains. *Icarus*, **154**, 40–79.

Clifford, S.M., Crisp, D., Fisher, D.A., *et al.* (2000). The state and future of Mars polar science and exploration. *Icarus*, **144**, 210–242.

Cockell, C.S. and Barlow, N.G. (2002). Impact excavation and the search for subsurface life on Mars. *Icarus*, **155**, 340–349.

Cockell, C.S., Catling, D.C., Davis, W.L., *et al.* (2000). The ultraviolet environment of Mars: biological implications past, present, and future. *Icarus*, **146**, 343–359.

Cohen, B.A., Swindle, T.D., and Kring, D.A. (2000). Support for the lunar cataclysm hypothesis from lunar meteorite impact melt ages. *Science*, **290**, 1754–1756.

Colaprete, A., Barnes, J.R., Haberle, R.M., *et al.* (2005). Albedo of the south pole of Mars determined by topographic forcing of atmospheric dynamics. *Nature*, **435**, 184–188.

Coleman, N.M. (2005). Martian megaflood triggered chaos formation, revealing groundwater depth, cryosphere thickness, and crustal heat flux. *Journal of Geophysical Research*, **110**, E12S20, doi: 10.1029/2005JE002419.

Collins, M., Lewis, S.R., Read, P.L., and Hourdin, F. (1996). Baroclinic wave transitions in the martian atmosphere. *Icarus*, **120**, 344–357.

Comer, R.P., Solomon, S.C., and Head, J.W. (1985). Thickness of the lithosphere from the tectonic response to volcanic loads. *Reviews of Geophysics*, **34**, 143–151.

Connerney, J.E.P., Acuña, M.H., Wasilewski, P.J., *et al.* (1999). Magnetic lineations in the ancient crust of Mars. *Science*, **284**, 794–798.

Connerney, J.E.P., Acuña, M.H., Ness, N.F., Spohn, T., and Schubert, G. (2004). Mars crustal magnetism. *Space Science Reviews*, **111**, 1–32.

Connerney, J.E.P., Acuña, M.H., Ness, N.F., *et al.* (2005). Tectonic implications of Mars crustal magnetism. *Proceedings of the National Academy of Sciences of the USA*, **102**, 14970–14975.

Conrath, B.J. (1981). Planetary-scale wave structure in the martian atmosphere. *Icarus*, **48**, 246–255.

COSPAR (2005). *COSPAR Planetary Protection Policy*. Available online at www.cosparhq.org/Scistr/PPPolicy.htm

Costard, F.M. and Kargel, J.S. (1995). Outwash plains and thermokarst on Mars. *Icarus*, **114**, 93–112.

Craddock, R.A. and Howard, A.D. (2002). The case for rainfall on a warm, wet early Mars. *Journal of Geophysical Research*, **107**, 5111, doi: 10.1029/2001JE001505.

Crater Analysis Techniques Working Group (1979). Standard techniques for presentation and analysis of crater size–frequency data. *Icarus*, **37**, 467–474.

Crown, D.A. and Greeley, R. (1993). Volcanic geology of Hadriaca Patera and the eastern Hellas region of Mars. *Journal of Geophysical Research*, **98**, 3431–3451.

Culler, T.S., Becker, T.A., Muller, R.A., and Renne, P.R. (2000). Lunar impact history from $^{40}$Ar/$^{39}$Ar dating of glass spherules. *Science*, **287**, 1785–1788.

Dalrymple, G.B. and Ryder, G. (1993). $^{40}$Ar/$^{39}$Ar age spectra of Apollo 15 impact melt rocks by laser step-heating and their bearing on the history of lunar basin formation. *Journal of Geophysical Research*, **98**, 13085–13095.

Dalrymple, G.B. and Ryder, G. (1996). Argon-40/argon-39 age spectra of Apollo 17 highlands breccia samples by laser step heating and the age of the Serenitatis basin. *Journal of Geophysical Research*, **101**, 26069–26084.

Davis, P.M. (1993). Meteoroid impacts as seismic sources on Mars. *Icarus,* **105**, 469–478.

deVaucouleurs, G., Davies, M.E., and Sturms, F.M. (1973). Mariner 9 aerographic coordinate system. *Journal of Geophysical Research*, **78**, 4395–4404.

DeVincenzi, D.L., Stabekis, P., and Barengoltz, J. (1996). Refinement of planetary protection policy for Mars missions. *Advances in Space Research*, **18**, 311–316.

Diaz, B. and Schulze-Makuch, D. (2006). Microbial survival rates of *Escherichia coli* and *Deinococcus radiodurans* under low temperature, low pressure, and UV-irradiation conditions, and their relevance to possible martian life. *Astrobiology*, **6**, 332–347.

Dohm, J.M. and Tanaka, K.L. (1999). Geology of the Thaumasia region, Mars: plateau development, valley origins, and magmatic evolution. *Planetary and Space Science*, **47**, 411–431.

Dollfus, A., Ebisawa, S., and Crussaire, D. (1996). Hoods, mists, frosts, and ice caps at the poles of Mars. *Journal of Geophysical Research*, **101**, 9207–9225.

Donahue, T.M. (1995). Evolution of water reservoirs on Mars from D/H ratios in the atmosphere and crust. *Nature*, **374**, 432–434.

Doran, P.T., Wharton, R.A., Des Marais, D.J., and McKay, C.P. (1998). Antarctic paleolake sediments and the search for extinct life on Mars. *Journal of Geophysical Research*, **103**, 28481–28493.

Drake, M.J. and Righter, K. (2002). Determining the composition of the Earth. *Nature*, **416**, 39–44.

Dreibus, G. and Wänke, H. (1985). Mars: a volatile-rich planet. *Meteoritics*, **20**, 367–382.

Dreibus, G. and Wänke, H. (1987). Volatiles on Earth and Mars: a comparison. *Icarus*, **71**, 225–240.

Drouart, A., Dubrulle, B., Gautier, D., and Robert, F. (1999). Structure and transport in the solar nebula from constraints on deuterium enrichment and giant planets formation. *Icarus*, **140**, 129–155.

Edgett, K.S. and Malin, M.C. (2000). New views of Mars eolian activity, materials, and surface properties: three vignettes from the Mars Global Surveyor Mars Orbiter Camera. *Journal of Geophysical Research*, **105**, 1623–1650.

Edgett, K.S., Butler, B.J., Zimbelman, J.R., and Hamilton, V.E. (1997). Geologic context of the Mars radar "Stealth" region in southwestern Tharsis. *Journal of Geophysical Research*, **102**, 21545–21567.

Elkins-Tanton, L.T. and Parmentier, E.M. (2006). Water and carbon dioxide in the martian magma ocean: early atmospheric growth, subsequent mantle compositions, and planetary cooling rates. *Lunar and Planetary Science XXXVII*, Abstract #2007 (CD-ROM). Houston, TX: Lunar and Planetary Institute.

Elkins-Tanton, L.T., Zaranek, S.E., Parmentier, E.M., and Hess, P.C. (2003). Early magnetic field and magmatic activity on Mars from magma ocean cumulate overturn. *Earth and Planetary Science Letters*, **236**, 1–12.

Elkins-Tanton, L.T., Hess, P.C., and Parmentier, E.M. (2005). Possible formation of ancient crust on Mars through magma ocean processes. *Journal of Geophysical Research*, **110**, E12S01, doi: 10.1029/2005JE002480.

Eluszkiewicz, J., Moncet, J.-L., Titus, T.N., and Hansen, G.B. (2005). A microphysically-based approach to modeling emissivity and albedo of the martian seasonal caps. *Icarus*, **174**, 524–534.

Erard, S. (2000). The 1994–1995 apparition of Mars observed from Pic-du-Midi. *Planetary and Space Science*, **48**, 1271–1287.

Fairén, A.G., Ruiz, J., and Anguita, F. (2002). An origin for the linear magnetic anomalies on Mars through accretion of terranes: implications for dynamo timing. *Icarus*, **160**, 220–223.

Farmer, J. (1998). Thermophiles, early biosphere evolution, and the origin of life on Earth: implications for the exobiological exploration of Mars. *Journal of Geophysical Research*, **103**, 28457–28461.

Farmer, J.D. and Des Marais, D.J. (1999). Exploring for a record of ancient martian life. *Journal of Geophysical Research*, **104**, 26977–26995.

Fasset, C.I. and Head, J.W. (2005). Fluvial sedimentary deposits on Mars: ancient deltas in a crater lake in the Nili Fossae region. *Geophysical Research Letters*, **32**, L14201, doi: 10.1029/2005GL023456.

Fasset, C.I. and Head, J.W. (2006). Valleys on Hecates Tholus, Mars: origin by basal melting of summit snowpack. *Planetary and Space Science*, **54**, 370–378.

Feldman, W.C., Boynton, W.V., Tokar, R.L., *et al.* (2002). Global distribution of neutrons from Mars: results from Mars Odyssey. *Science*, **297**, 75–78.

Feldman, W.C., Prettyman, T.H., Maurice, S., *et al.* (2004a). Global distribution of near-surface hydrogen on Mars. *Journal of Geophysical Research*, **109**, E09006, doi: 10.1029/2003JE002160.

Feldman, W.C., Mellon, M.T., Maurice, S., *et al.* (2004b). Hydrated states of $MgSO_4$ at equatorial latitudes on Mars. *Geophysical Research Letters*, **31**, L16702, doi: 10.1029/2004GL020181.

Ferguson, D.C., Kolecki, J.C., Siebert, M.W., Wilt, D.M., and Matijevic, J.R. (1999). Evidence for martian electrostatic charging and abrasive wheel wear from the Wheel Abrasion Experiment on the Pathfinder Sojourner rover. *Journal of Geophysical Research*, **104**, 8747–8789.

Fernández-Remolar, D.C., Morris, R.V., Gruener, J.E., Amils, R., and Knoll, A.H. (2005). The Rio Tinto Basin, Spain: mineralogy, sedimentary geobiology, and implications for interpretation of outcrop rocks at Meridiani Planum, Mars. *Earth and Planetary Science Letters*, **240**, 149–167.

Fialips, C.I., Carey, J.W., Vaniman, D.T., *et al.* (2005). Hydration states of zeolites, clays, and hydrated salts under present-day martian surface conditions: can hydrous minerals account for Mars Odyssey observations of near-equatorial water-equivalent hydrogen? *Icarus*, **178**, 74–83.

Fishbaugh, K.E. and Head, J.W. (2000). North polar region of Mars: topography of circumpolar deposits from Mars Orbiter Laser Altimeter (MOLA) data and evidence for asymmetric retreat of the polar cap. *Journal of Geophysical Research*, **105**, 22433–22486.

Fishbaugh, K.E. and Head, J.W. (2002). Chasma Boreale, Mars: topographic characterization from Mars Orbiter Laser Altimeter data and implications for mechanisms of formation. *Journal of Geophysical Research*, **107**, 5013, doi: 10.1029/2000JE001351.

Fisher, D.A. (1993). If martian ice caps flow: ablation mechanisms and appearance. *Icarus*, **105**, 501–511.

Fisher, D.A. (2000). Internal layers in an "accublation" ice cap: a test for flow. *Icarus*, **144**, 289–294.

Fisher, J.A., Richardson, M.I., Newman, C.E., *et al.* (2005). A survey of martian dust devil activity using Mars Global Surveyor Mars Orbiter Camera images. *Journal of Geophysical Research*, **110**, E03004, doi: 10.1029/2003JE002165.

Folkner, W.M., Yoder, C.F., Yuan, D.N., Standish, E.M., and Preston, R.A. (1997). Interior structure and seasonal mass redistribution of Mars from radio tracking of Mars Pathfinder. *Science*, **278**, 1749–1752.

Forget, F. and Pierrehumbert, R.T. (1997). Warming early Mars with carbon dioxide clouds that scatter infrared radiation. *Science*, **278**, 1273–1276.

Forget, F., Hourdin, F., Fournier, R., *et al.* (1999). Improved general circulation models of the martian atmosphere from the surface to above 80 km. *Journal of Geophysical Research*, **104**, 24155–24176.

Formisano, V., Atreya, S., Encrenaz, T., Ignatiev, N., and Giuranna, M. (2004). Detection of methane in the atmosphere of Mars. *Science*, **306**, 1758–1761.

Forsberg-Taylor, N.K., Howard, A.D., and Craddock, R.A. (2004). Crater degradation in the martian highlands: morphometric analysis of the Sinus Sabaeus region and simulation modeling suggest fluvial processes. *Journal of Geophysical Research*, **109**, E05002, doi: 10.1029/2004JE002242.

Frahm, R.A., Winningham, J.D., Sharber, J.R., *et al.* (2006). Carbon dioxide photoelectron energy peaks at Mars. *Icarus*, **182**, 371–382.

French, B.M. (1998). *Traces of Catastrophe: A Handbook of Shock-Metamorphic Effects in Terrestrial Meteorite Impact Structures*, Contribution No. 954. Houston, TX: Lunar and Planetary Institute.

Frey, H. and Schultz, R.A. (1988). Large impact basins and the mega-impact origin for the crustal dichotomy on Mars. *Geophysical Research Letters*, **15**, 229–232.

Frey, H.V., Roark, J.H., Shockey, K.M., Frey, E.L., and Sakimoto, S.E.H. (2002). Ancient lowlands on Mars. *Geophysical Research Letters*, **29**, 1384, doi: 10.1029/2001GL013832.

Fukuhara, T. and Imamura, T. (2005). Waves encircling the summer southern pole of Mars observed by MGS TES. *Geophysical Research Letters*, **32**, L18811, doi: 10.1029/2005GL023819.

Garvin, J.B. and Frawley, J.J. (1998). Geometric properties of martian impact craters: preliminary results from the Mars Orbiter Laser Altimeter. *Geophysical Research Letters*, **25**, 4405–4408.

Garvin, J.B., Sakimoto, S.E.H., Frawley, J.J., and Schnetzler, C. (2000a). North polar region craterforms on Mars: geometric characteristics from the Mars Orbiter Laser Altimeter. *Icarus*, **144**, 329–352.

Garvin, J.B., Sakimoto, S.E.H., Frawley, J.J., Schnetzler, C.C., and Wright, H.M. (2002b). Topographic evidence for geologically recent near-polar volcanism on Mars. *Icarus*, **145**, 648–652.

Garvin, J.B., Sakimoto, S.E.H., and Frawley, J.J. (2003). Craters on Mars: global geometric properties from gridded MOLA topography. In *6th International Conference on Mars*, Abstract #3277. Houston, TX: Lunar and Planetary Institute.

Gault, D.E., Quaide, W.L., and Oberbeck, V.R. (1968). Impact cratering mechanics and structures. In *Shock Metamorphism of Natural Materials*, ed. B.M. French and N.M. Short. Baltimore, MD: Mono Book Corporation, pp.87–99.

Geiss, J. and Gloeckler, G. (1998). Abundances of deuterium and helium-3 in the protosolar cloud. *Space Science Reviews*, **84**, 239–250.

Gellert, R., Rieder, R., Anderson, R.C., *et al.* (2004). Chemistry of rocks and soils in Gusev Crater from the alpha particle X-ray spectrometer. *Science*, **305**, 829–836.

Gellert, R., Rieder, R., Brückner, J., *et al.* (2006). Alpha Particle X-Ray Spectrometer (APSX): results from Gusev Crater and calibration report. *Journal of Geophysical Research*, **111**, E02S05, doi: 10.1029/2005JE002555.

Gendrin, A., Mangold, N., Bibring, J.-P., *et al.* (2005). Sulfates in martian layered terrains: the OMEGA/Mars Express view. *Science*, **307**, 1587–1591.

Ghatan, G.J. and Head, J.W. (2002). Candidate subglacial volcanoes in the south polar region of Mars: morphology, morphometry, and eruption conditions. *Journal of Geophysical Research*, **107**, 5048, doi: 10.1029/2001JE001519.

Glaze, L.S., Baloga, S.M., and Stofan, E.R. (2003). A methodology for constraining lava flow rheologies with MOLA. *Icarus*, **165**, 26–33.

Golden, D.C., Ming, D.W., Morris, R.V., and Mertzman, S.A. (2005). Laboratory-simulated acid-sulfate weathering of basaltic materials: implications for formation of sulfates at Meridiani Planum and Gusev Crater, Mars. *Journal of Geophysical Research*, **110**, E12S07, doi: 10.1029/2005JE002451.

Golombek, M.P. and Rapp, D. (1997). Size–frequency distributions of rocks on Mars and Earth analog sites: implications for future landed missions. *Journal of Geophysical Research*, **102**, 4117–4129.

Golombek, M.P., Banerdt, W.B., Tanaka, K.L., and Tralli, D.M. (1992). A prediction of Mars seismicity from surface faulting. *Science*, **258**, 979–981.

Golombek, M.P., Anderson, R.C., Barnes, J.R., *et al.* (1999a). Overview of the Mars Pathfinder Mission: launch through landing, surface operations, data sets, and science results. *Journal of Geophysical Research*, **104**, 8523–8553.

Golombek, M.P., Moore, H.J., Haldemann, A.F.C., Parker, T.J., and Schofield, J.T. (1999b). Assessment of Mars Pathfinder landing site predictions. *Journal of Geophysical Research*, **104**, 8585–8594.

Golombek, M.P., Anderson, F.S., and Zuber, M.T. (2001). Martian wrinkle ridge topography: evidence for subsurface faults from MOLA. *Journal of Geophysical Research*, **106**, 23811–23821.

Golombek, M.P., Grant, J.A., Parker, T.J., *et al.* (2003). Selection of the Mars Exploration Rover landing sites. *Journal of Geophysical Research*, **108**, 8072, doi: 10.1029/2003JE002074.

Golombek, M.P., Arvidson, R.E., Bell, J.F., *et al.* (2005). Assessment of Mars Exploration Rover landing site predictions. *Nature*, **436**, 44–48.

Golombek, M.P., Crumpler, L.S., Grant, J.A., *et al.* (2006). Geology of the Gusev cratered plains from the Spirit rover traverse. *Journal of Geophysical Research*, **111**, E02S07, doi: 10.1029/2005JE002503.

Gomes, R., Levison, H.F., Tsiganis, K., and Morbidelli, A. (2005). Origin of the cataclysmic Late Heavy Bombardment period of the terrestrial planets. *Nature*, **435**, 466–469.

Goudy, C.L., Schultz, R.A., and Gregg, T.K.P. (2005). Coulomb stress changes in Hesperia Planum, Mars, reveal regional thrust fault reactivation. *Journal of Geophysical Research*, **110**, E10005, doi: 10.1029/2004JE02293.

Grant, J.A. and Parker, T.J. (2002). Drainage evolution in the Margaritifer Sinus region, Mars. *Journal of Geophysical Research*, **107**, 5066, doi: 10.1029/2001JE001678.

Greeley, R. and Fagents, S.A. (2001). Icelandic pseudocraters as analogs to some volcanic cones on Mars. *Journal of Geophysical Research*, **106**, 20527–20546.

Greeley, R. and Iverson, J.D. (1985). *Wind as a Geological Process on Earth, Mars, Venus, and Titan.* Cambridge, UK: Cambridge University Press.

Greeley, R., Kraft, M., Sullivan, R., *et al.* (1999). Aeolian features and processes at the Mars Pathfinder landing site. *Journal of Geophysical Research*, **104**, 8573–8584.

Greeley, R., Arvidson, R.E., Barlett, P.W., *et al.* (2006). Gusev Crater: wind-related features and processes observed by the Mars Exploration Rover Spirit. *Journal of Geophysical Research*, **111**, E02S09, doi: 10.1029/2005JE002491.

Gregg, T.K.P. and Williams, S.N. (1996). Explosive mafic volcanoes on Mars and Earth: deep magma sources and rapid rise rate. *Icarus*, **122**, 397–405.

Grossman, L. (1972). Condensation in the primitive solar nebula. *Geochimica et Cosmochimica Acta*, **36**, 597–619.

Grott, M., Hauber, E., Werner, S.C., Kronberg, P., and Neukum, G. (2005). High heat flux on ancient Mars: evidence from rift flank uplift at Coracis Fossae. *Geophysical Research Letters*, **32**, L21201, doi: 10.1029/2005GL023894.

Gulick, V.C. (1998). Magmatic intrusions and a hydrothermal origin for fluvial valleys on Mars. *Journal of Geophysical Research*, **103**, 19 365–19 388.

Gulick, V.C. and Baker, V.R. (1990). Origin and evolution of valleys on martian volcanoes. *Journal of Geophysical Research*, **95**, 14 325–14 344.

Gulick, V.C., Tyler, D., McKay, C.P., and Haberle, R.M. (1997). Episodic ocean-induced $CO_2$ greenhouse on Mars: implications for fluvial valley formation. *Icarus*, **130**, 68–86.

Haberle, R.M. (1998). Early Mars climate models. *Journal of Geophysical Research*, **103**, 28 467–28 479.

Haberle, R.M., Pollack, J.B., Barnes, J.R., *et al.* (1993). Mars atmospheric dynamics as simulated by the NASA/Ames general circulation model. 1. The zonal-mean circulation. *Journal of Geophysical Research*, **98**, 3093–3123.

Haberle, R.M., McKay, C.P., Schaeffer, J., *et al.* (2001). On the possibility of liquid water on present-day Mars. *Journal of Geophysical Research*, **106**, 23 317–23 326.

Haberle, R.M., Murphy, J.R., and Schaeffer, J. (2003). Orbital change experiments with a Mars general circulation model. *Icarus*, **161**, 66–89.

Halliday, A.N., Wänke, H., Birck, J.-L., and Clayton, R.N. (2001). The accretion, composition, and early differentiation of Mars. *Space Science Reviews*, **96**, 197–230.

Hamilton, V.E., Christensen, P.R., McSween, H.Y., and Bandfield, J.L. (2003). Searching for the source regions of martian meteorites using MGS TES: integrating martian meteorites into the global distribution of igneous materials on Mars. *Meteoritics and Planetary Science*, **38**, 871–885.

Hanel, R.A., Conrath, B.J., Jennings, D.E., and Samuelson, R.E. (2003). *Exploration of the Solar System by Infrared Remote Sensing*, 2nd edn. Cambridge, UK: Cambridge University Press.

Hargraves, R.B., Collinson, D.W., Arvidson, R.E., and Spitzer, C.R. (1977). The Viking magnetic properties experiment: primary mission results. *Journal of Geophysical Research*, **82**, 4547–4558.

Hargraves, R.B., Collinson, D.W., Arvidson, R.E., and Cates, P.M. (1979). The Viking magnetic properties experiment: extended mission results. *Journal of Geophysical Research*, **84**, 8379–8384.

Harmon, J.K., Arvidson, R.E., Guinness, E.A., Campbell, B.A., and Slade, M.A. (1999). Mars mapping with delay-Doppler radar. *Journal of Geophysical Research*, **104**, 14065–14089.

Harper, C.L., Nyquist, L.E., Bansal, B., Weismann, H., and Shih, C.-Y. (1995). Rapid accretion and early differentiation of Mars indicated by $^{142}Nd/^{144}Nd$ in SNC meteorites. *Science*, **267**, 213–217.

Harrison, K.P. and Grimm, R.E. (2003). Rheological constraints on martian landslides. *Icarus*, **163**, 347–362.

Hartmann, W.K. (1984). Does crater "saturation equilibrium" occur in the solar system? *Icarus*, **60**, 56–74.

Hartmann, W.K. (1990). Additional evidence about an early intense flux of asteroids and the origin of Phobos. *Icarus*, **87**, 236–240.

Hartmann, W.K. (1997). Planetary cratering. 2. Studies of saturation equilibrium. *Meteoritics*, **32**, 109–121.

Hartmann, W.K. (2003). Megaregolith evolution and cratering cataclysm models: lunar cataclysm as a misconception (28 years later). *Meteoritics and Planetary Science*, **38**, 579–593.

Hartmann, W.K. (2005). Martian cratering. 8. Isochron refinement and the chronology of Mars. *Icarus*, **174**, 294–320.

Hartmann, W.K. and Barlow, N.G. (2006). Nature of the martian uplands: effects on martian meteorite age distribution and secondary cratering. *Meteoritics and Planetary Science*, **41**, 1453–1467.

Hartmann, W.K. and Berman, D.C. (2000). Elysium Planitia lava flows: crater count chronology and geological implications. *Journal of Geophysical Research*, **105**, 15011–15025.

Hartmann, W.K. and Davis, D.R. (1975). Satellite-sized planetesimals and lunar origin. *Icarus*, **24**, 504–515.

Hartmann, W.K. and Neukum, G. (2001). Cratering chronology and the evolution of Mars. *Space Science Reviews*, **96**, 165–194.

Hartmann, W.K., Malin, M., McEwen, A., *et al.* (1999). Evidence for recent volcanism on Mars from crater counts. *Nature*, **397**, 586–589.

Harvey, R.P. and Hamilton, V.E. (2005). Syrtis Major as the source region of the Nakhlite/Chassigny group of martian meteorites: implications for the geological history of Mars. *Lunar and Planetary Science XXXVI*, Abstract #1019 (CD-ROM). Houston, TX: Lunar and Planetary Institute.

Hauber, E., van Gasselt, S., Ivanov, B., *et al.* (2005). Discovery of a flank caldera and very young glacial activity at Hecates Tholus, Mars. *Nature*, **434**, 356–361.

Head, J.N., Melosh, H.J., and Ivanov, B.A. (2002). Martian meteorite launch: high-speed ejecta from small craters. *Science*, **298**, 1752–1756.

Head, J.W. and Marchant, D.R. (2003). Cold-based mountain glaciers on Mars: western Arsia Mons. *Geology*, **31**, 641–644.

Head, J.W. and Pratt, S. (2001). Extensive Hesperian-aged south polar ice cap on Mars: evidence for massive melting and retreat, and lateral flow and ponding of meltwater. *Journal of Geophysical Research*, **106**, 12275–12299.

Head, J.W., Kreslavsky, M., Hiesinger, H., *et al.* (1998). Oceans in the past history of Mars: tests for their presence using Mars Orbiter Laser Altimeter (MOLA) data. *Geophysical Research Letters*, **25**, 4401–4404.

Head, J.W., Hiesinger, H., Ivanov, M.A., *et al.* (1999). Possible ancient oceans on Mars: evidence from Mars Orbiter Laser Altimeter data. *Science*, **286**, 2134–2137.

Head, J.W., Kreslavsky, M.A., and Pratt, S. (2002). Northern lowlands of Mars: evidence for widespread volcanic flooding and tectonic deformation in the Hesperian Period. *Journal of Geophysical Research*, **107**, 5003, doi: 10.1029/2000JE001445.

Head, J.W., Mustard, J.F., Kreslavsky, M.A., Milliken, R.E., and Marchant, D.R. (2003a). Recent ice ages on Mars. *Nature*, **426**, 797–802.

Head, J.W., Wilson, L., and Mitchell, K.L. (2003b). Generation of recent massive water floods at Cerberus Fossae, Mars by dike emplacement, cryospheric cracking, and confined aquifer groundwater release. *Geophysical Research Letters*, **30**, 1577, doi: 10.1029/2003GL017135.

Head, J.W., Neukum, G., Jaumann, R., *et al.* (2005). Tropical to mid-latitude snow and ice accumulation, flow and glaciation on Mars. *Nature*, **434**, 346–351.

Head, J.W., Marchant, D.R., Agnew, M.C., Fassett, C.I., and Kreslavsky, M.A. (2006a). Extensive valley glacier deposits in the northern mid-latitudes of Mars: evidence for Late Amazonian obliquity-driven climate change. *Earth and Planetary Science Letters*, **241**, 663–671.

Head, J.W., Nahm, A.L., Marchant, D.R., and Neukum, G. (2006b). Modification of the dichotomy boundary on Mars by Amazonian mid-latitude regional glaciation. *Geophysical Research Letters*, **33**, L08S03, doi: 10.1029/2005GL024360.

Hecht, M.H. (2002). Metastability of liquid water on Mars. *Icarus*, **156**, 373–386.

Heiken, G.H., Vaniman, D.T., and French, B.M. (1991). *Lunar Sourcebook: A User's Guide to the Moon.* Cambridge, UK: Cambridge University Press.

Heldmann, J.L., Toon, O.B., Pollard, W.H., *et al.* (2005). Formation of martian gullies by the actions of liquid water flowing under current martian environmental conditions. *Journal of Geophysical Research*, **110**, E05004, doi: 10.1029/2004JE002261.

Herkenhoff, K.E. and Plaut, J.J. (2000). Surface ages and resurfacing rates of the polar layered deposits, Mars. *Icarus*, **144**, 243–253.

Herkenhoff, K.E. and Vasavada, A.R. (1999). Dark material in the polar layered deposits and dunes on Mars. *Journal of Geophysical Research*, **104**, 16487–16500.

Herkenhoff, K.E., Squyres, S.W., Arvidson, R., *et al.* (2004). Evidence from Opportunity's Microscopic Imager for water on Meridiani Planum. *Science*, **306**, 1727–1730.

Hiesinger, H. and Head, J.W. (2000). Characteristics and origin of polygonal terrain in southern Utopia Planitia, Mars: results from Mars Orbiter Laser Altimeter and Mars Orbiter Camera data. *Journal of Geophysical Research*, **105**, 11999–12022.

Hiesinger, H. and Head, J.W. (2002). Topography and morphology of the Argyre Basin, Mars: implications for its geologic and hydrologic history. *Planetary and Space Science*, **50**, 939–981.

Hiesinger, H. and Head, J.W. (2004). The Syrtis Major volcanic province, Mars: synthesis from Mars Global Surveyor data. *Journal of Geophysical Research*, **109**, E01004, doi: 10.1029/2003JE002143.

Hinson, D.P. and Wilson, R.J. (2002). Transient eddies in the southern hemisphere of Mars. *Geophysical Research Letters*, **29**, 1154, doi: 10.1029/2001GL014103.

Hinson, D.P., Simpson, R.A., Twicken, J.D., Tyler, G.L., and Flasar, F.M. (1999). Initial results from radio occultation measurements with Mars Global Surveyor. *Journal of Geophysical Research*, **104**, 26997–27012.

Hinson, D.P., Tyler, G.L., Hollingsworth, J.L., and Wilson, R.J. (2001). Radio occultation measurements of forced atmospheric waves on Mars. *Journal of Geophysical Research*, **106**, 1463–1480.

Hinson, D.P., Wilson, R.J., Smith, M.D., and Conrath, B.J. (2003). Stationary planetary waves in the atmosphere of Mars during southern winter. *Journal of Geophysical Research*, **108**, 5004, doi: 10.1029/2002JE001949.

Hodges, C.A. and Moore, H.J. (1994). *Atlas of Volcanic Landforms on Mars*, Professional Paper No. 1534. Washington, DC: US Geological Survey.

Hoefen, T.M., Clark, R.N., Bandfield, J.L., *et al.* (2003). Discovery of olivine in the Nili Fossae region of Mars. *Science*, **302**, 627–630.

Hoffman, N. (2000). White Mars: a new model for Mars' surface and atmosphere based on $CO_2$. *Icarus*, **146**, 326–342.

Hollingsworth, J.L., Haberle, R.M., Bridger, A.F.C., *et al.* (1996). Winter storm zones on Mars. *Nature*, **380**, 413–416.

Hood, L.L. and Zakharian, A. (2001). Mapping and modeling of magnetic anomalies in the northern polar region of Mars. *Journal of Geophysical Research*, **106**, 14601–14619.

Hood, L.L., Richmond, N.C., Pierazzo, E., and Rochette, P. (2003). Distribution of crustal magnetic fields on Mars: shock effects of basin-forming impacts. *Geophysical Research Letters*, **30**, 1281, doi: 10.1029/2002GL016657.

Horowitz, N.H., Hobby, G.L., and Hubbard, J.S. (1977). Viking on Mars: the carbon assimilation experiments. *Journal of Geophysical Research*, **82**, 4659–4662.

Houben, H., Haberle, R.M., Young, R.E., and Zent, A.P. (1997). Modeling the martian seasonal water cycle. *Journal of Geophysical Research*, **102**, 9069–9083.

Hourdin, F., Forget, F., and Talagrand, O. (1995). The sensitivity of the martian surface pressure to various parameters: a comparison between numerical simulations and Viking observations. *Journal of Geophysical Research*, **100**, 5501–5523.

Howard, A.D. (2000). The role of eolian processes in forming surface features of the martian polar layered deposits. *Icarus*, **144**, 267–288.

Howard, A.D., Cutts, J.A., and Blasius, K.R. (1982). Stratigraphic relationships within martian polar cap deposits. *Icarus*, **50**, 161–215.

Howard, A.D., Moore, J.M., and Irwin, R.P. (2005). An intense terminal epoch of widespread fluvial activity on early Mars. 1. Valley network incision and associated deposits. *Journal of Geophysical Research*, **110**, E12S14, doi: 10.1029/2005JE002459.

Hubbard, W.B. (1984). *Planetary Interiors*. New York: Van Nostrand Reinhold.

Hunten, D.M., Pepin, R.O., and Walker, J.G.C. (1987). Mass fractionation in hydrodynamic escape. *Icarus*, **69**, 532–549.

Hviid, S.F., Madsen, M.B., Gunnlaugsson, H.P., et al. (1997). Magnetic properties experiments on the Mars Pathfinder Lander: preliminary results. *Science*, **278**, 1997.

Hynek, B.M., Phillips, R.J., and Arvidson, R.E. (2003). Explosive volcanism in the Tharsis region: global evidence in the martian geologic record. *Journal of Geophysical Research*, **108**, 5111, doi: 10.1029/2003JE002062.

Irwin, R.P. and Howard, A.D. (2002). Drainage basin evolution in Noachian Terra Cimmeria, Mars. *Journal of Geophysical Research*, **107**, 5056, doi: 10.1029/2001JE001818.

Irwin, R.P., Watters, T.R., Howard, A.D., and Zimbelman, J.R. (2004). Sedimentary, resurfacing and fretted terrain development along the crustal dichotomy boundary, Aeolis Mensae, Mars. *Journal of Geophysical Research*, **109**, E09011, doi: 10.1029/2004JE002248.

Irwin, R.P., Craddock, R.A., and Howard, A.D. (2005). Interior channels in martian valley networks: discharge and runoff production. *Geology*, **33**, 489–492.

Ivanov, B.A. (2001). Mars/Moon cratering rate ratio estimates. *Space Science Reviews*, **96**, 87–104.

Ivanov, B.A. (2006). Cratering rate comparisons between terrestrial planets. In *Workshop on Surface Ages and Histories: Issues in Planetary Chronology*, Contribution No. 1320. Houston, TX: Lunar and Planetary Institute, pp. 26–27.

Ives, H.E. (1919). Some large-scale experiments imitating the craters of the Moon. *Astrophysical Journal*, **50**, 245.

Jagoutz, E., Sorowka, A., Vogel, J.D., and Wänke, H. (1994). ALH84001: alien or progenitor of the SNC family? *Meteoritics*, **28**, 478–479.

Jakosky, B.M. and Haberle, R.M. (1992). The seasonal behavior of water on Mars. In *Mars*, ed. H.H. Kieffer, B.M. Jakosky, C.W. Snyder, and M.S. Matthews. Tucson, AZ: University of Arizona Press, pp. 969–1016.

Jakosky, B.M. and Jones, J.H. (1997). The history of martian volatiles. *Reviews of Geophysics*, **35**, 1–16.

Jakosky, B.M. and Phillips, R.J. (2001). Mars' volatile and climate history. *Nature*, **412**, 237–244.

Jakosky, B.M., Henderson, B.G., and Mellon, M.T. (1993). The Mars water cycle at other epochs: recent history of the polar caps and layered terrain. *Icarus*, **102**, 286–297.

Jakosky, B.M., Pepin, R.O., Johnson, R.E., and Fox, J.L. (1994). Mars atmospheric loss and isotopic fractionation by solar-wind-induced sputtering and photochemical escape. *Icarus*, **111**, 271–288.

Jakosky, B.M., Mellon, M.T., Kieffer, H.H., *et al.* (2000). The thermal inertia of Mars from the Mars Global Surveyor Thermal Emission Spectrometer. *Journal of Geophysical Research*, **105**, 9643–9652.

Jakosky, B.M., Nealson, K.H., Bakermans, C., *et al.* (2003). Subfreezing activity of microorganisms and the potential habitability of Mars' polar regions. *Astrobiology*, **3**, 343–350.

Jakosky, B.M., Mellon, M.T., Varnes, E.S., *et al.* (2005). Mars low-latitude neutron distribution: possible remnant near-surface water ice and a mechanism for its recent emplacement. *Icarus*, **175**, 58–67.

James, P.B. and Cantor, B.A. (2002). Atmospheric monitoring of Mars by the Mars Orbiter Camera on Mars Global Surveyor. *Advances in Space Research*, **29**, 121–129.

James, P.B., Kieffer, H.H., and Paige, D.A. (1992). The seasonal cycle of carbon dioxide on Mars. In *Mars,* ed. H.H. Kieffer, B.M. Jakosky, C.W. Snyder, and M.S. Matthews. Tucson, AZ: University of Arizona Press, pp. 934–968.

James, P.B., Bell, J.F., Clancy, R.T., *et al.* (1996). Global imaging of Mars by Hubble Space Telescope during the 1995 opposition. *Journal of Geophysical Research*, **101**, 18 883–18 890.

James, P.B., Hansen, G.B., and Titus, T.N. (2005). The carbon dioxide cycle. *Advances in Space Research*, **35**, 14–20.

Johnston, D.H., McGetchin, T.R., and Toksöz, M.N. (1974). The thermal state and internal structure of Mars. *Journal of Geophysical Research*, **79**, 3959–3971.

Joshi, M.M., Lewis, S.R., Read, P.L., and Catling, D.C. (1995). Western boundary currents in the martian atmosphere: numerical simulations and observational evidence. *Journal of Geophysical Research*, **100**, 5485–5500.

Joshi, M.M., Haberle, R.M., Barnes, J.R., Murphy, J.R., and Schaeffer, J. (1997). Low-level jets in the NASA Ames Mars general circulation model. *Journal of Geophysical Research*, **102**, 6511–6523.

Kahn, R.A., Martin, T.Z., Zurek, R.W., and Lee, S.W. (1992). The martian dust cycle. In *Mars*, ed. H.H. Kieffer, B.M. Jakosky, C.W. Snyder, and M.S. Matthews. Tucson, AZ: University of Arizona Press, pp. 1017–1053.

Kahre, M.A., Murphy, J.R., and Haberle, R.M. (2006). Modeling the martian dust cycle and surface dust reservoirs with the NASA Ames general circulation model. *Journal of Geophysical Research*, **111**, E06008, doi: 10.1029/2005JE002588.

Kargel, J.S. and Strom, R.G. (1992). Ancient glaciation on Mars. *Geology*, **20**, 3–7.

Kargel, J.S., Baker, V.R., Begét, J.E., *et al.* (1995). Evidence of continental glaciation in the martian northern plains. *Journal of Geophysical Research*, **100**, 5351–5368.

Karlsson, H.R., Clayton, R.N., Gibson, E.K., and Mayeda, T.K. (1992). Water in SNC meteorites: evidence for a martian hydrosphere. *Science*, **255**, 1409–1411.

Kass, D.M., Schofield, J.T., Michaels, T.I., *et al.* (2003). Analysis of atmospheric mesoscale models for entry, descent, and landing. *Journal of Geophysical Research*, **108**, 8090, doi: 10.1029/2003JE002065.

Kasting, J.F. (1991). $CO_2$ condensation and the climate of early Mars. *Icarus*, **94**, 1–13.

Keszthelyi, L., McEwen, A.S., and Thordarson, T. (2000). Terrestrial analogs and thermal models for martian flood lavas. *Journal of Geophysical Research*, **105**, 15 027–15 049.

Kiefer, W.S. (2003). Melting in the martian mantle: shergottite formation and implications for present-day mantle convection on Mars. *Meteoritics and Planetary Science*, **39**, 1815–1832.

Kieffer, H.H. and Titus, T.N. (2001). TES mapping of Mars' north seasonal cap. *Icarus*, **154**, 162–180.

Kieffer, H.H., Jakosky, B.M., and Snyder, C.W. (1992). The planet Mars: from antiquity to the present. In *Mars*, ed. H.H. Kieffer, B.M. Jakosky, C.W. Snyder, and M.S. Matthews. Tucson, AZ: University of Arizona Press, pp. 1–33.

Kieffer, H.H., Titus, T.N., Mullins, K.F., and Christensen, P.R. (2000). Mars south polar spring and summer behavior observed by TES: seasonal cap evolution controlled by frost grain. *Journal of Geophysical Research*, **105**, 9653–9699.

Kieffer, H.H., Christensen, P.R., and Titus, T.N. (2006). $CO_2$ jets formed by sublimation beneath translucent slab ice in Mars' seasonal south polar ice cap. *Nature*, **442**, 793–796.

Klein, H.P. (1977). The Viking biological investigation: general aspects. *Journal of Geophysical Research*, **82**, 4677–4680.

Klein, H.P. (1998). The search for life on Mars: what we learned from Viking. *Journal of Geophysical Research*, **103**, 28463–28466.

Kleine, T., Münker, C., Mezger, K., and Palme, H. (2002). Rapid accretion and early core formation on asteroids and the terrestrial planets from Hf–W chronometry. *Nature*, **418**, 952–955.

Kletetschka, G., Connerney, J.E.P., Ness, N.F., and Acuña, M.H. (2004). Pressure effects on martian crustal magnetization near large impact basins. *Meteoritics and Planetary Science*, **39**, 1839–1848.

Klingelhöfer, G., Morris, R.V., Bernhardt, B., *et al.* (2003). Athena MIMOS II Mössbauer spectrometer investigation. *Journal of Geophysical Research*, **108**, 8067, doi: 10.1029/2003JE002138.

Klingelhöfer, G., Morris, R.V., Bernhardt, B., *et al.* (2004). Jarosite and hematite at Meridiani Planum from Opportunity's Mössbauer spectrometer. *Science*, **306**, 1740–1745.

Knoll, A.H., Carr, M., Clark, B., *et al.* (2005). An astrobiological perspective on Meridiani Planum. *Earth and Planetary Science Letters*, **240**, 179–189.

Kokubo, E. and Ida, S. (1998). Oligarchic growth of protoplanets. *Icarus*, **131**, 171–178.

Kokubo, E., Canup, R.M., and Ida, S. (2000). Lunar accretion from an impact-generated disk. In *Origin of the Earth and Moon*, ed. R.M. Canup and K. Righter. Tucson, AZ: University of Arizona Press, pp. 145–163.

Kolb, E.J. and Tanaka, K.L. (2001). Geologic history of the polar regions of Mars based on Mars Global Surveyor data. 2. Amazonian Period. *Icarus*, **153**, 22–39.

Kortenkamp, S.J., Kokubo, E., and Weidenschilling, S.J. (2000). Formation of planetary embryos. In *Origin of the Earth and Moon*, ed. R.M. Canup and K. Righter. Tucson, AZ: University of Arizona Press, pp. 85–100.

Kossacki, K.J., Markiewicz, W.J., and Smith, M.D. (2003). Surface temperature of martian regolith with polygonal features: influence of the subsurface water ice. *Planetary and Space Science*, **51**, 569–580.

Krasnopolsky, V. (2000). On the deuterium abundance on Mars and some related problems. *Icarus*, **148**, 597–602.

Krasnopolsky, V.A. (2006). Some problems related to the origin of methane on Mars. *Icarus*, **180**, 359–367.

Krasnopolsky, V.A., Maillard, J.P., and Owen, T.C. (2004). Detection of methane in the martian atmosphere: evidence for life? *Icarus*, **172**, 537–547.

Krauss, C.E., Horányi, M., and Robertson, S. (2006). Modeling the formation of electrostatic discharges on Mars. *Journal of Geophysical Research*, **111**, E02001, doi: 10.1029/2004JE002313.

Kreslavsky, M.A. and Head, J.W. (2000). Kilometer-scale roughness of Mars' surface: results from MOLA data analysis. *Journal of Geophysical Research*, **105**, 26695–26712.

Kress, M.E. and McKay, C.P. (2004). Formation of methane in comet impacts: implications for Earth, Mars, and Titan. *Icarus*, **168**, 475–483.

Kring, D.A. and Cohen, B.A. (2002). Cataclysmic bombardment throughout the inner solar system 3.9–4.0 Ga. *Journal of Geophysical Research*, **107**, 5009, doi: 10.1029/2001JE001529.

Kuiper, G.P., Whitaker, E.A., Strom, R.G., Fountain, J.W., and Larson, S.M. (1967). *Consolidated Lunar Atlas*. Tucson, AZ: Lunar and Planetary Laboratory, University of Arizona.

Kuzmin, R.O., Zabalueva, E.V., Mitrofanov, I.G., *et al.* (2004). Regions of potential existence of free water (ice) in the near-surface martian ground: results from the Mars Odyssey High-Energy Neutron Detector (HEND). *Solar System Research*, **38**, 1–11.

Lamb, M.P., Howard, A.D., Johnson, J., *et al.* (2006). Can springs cut canyons into rock? *Journal of Geophysical Research*, **111**, E07002, doi: 10.1029/2005JE002663.

Lanagan, P.D., McEwen, A.S., Keszthelyi, L.P., and Thordarson, T. (2001). Rootless cones on Mars indicating the presence of shallow equatorial ground ice in recent times. *Geophysical Research Letters,* **28**, 2365–2368.

Langevin, Y., Poulet, F., Bibring, J.-P., and Gondet, B. (2005a). Sulfates in the north polar region of Mars detected by OMEGA/Mars Express. *Science*, **307**, 1584–1586.

Langevin, Y., Poulet, F., Bibring, J.-P., *et al.* (2005b). Summer evolution of the north polar cap of Mars as observed by OMEGA/Mars Express. *Science*, **307**, 1581–1584.

Langlais, B., Purucker, M.E., and Mandea, M. (2004). Crustal magnetic field of Mars. *Journal of Geophysical Research*, **109**, E02008, doi: 10.1029/2003JE002048.

Laskar, J. and Robutel, P. (1993). The chaotic obliquity of the planets. *Nature*, **361**, 608–612.

Laskar, J., Levrard, B., and Mustard, J.F. (2002). Orbital forcing of the martian polar layered deposits. *Nature*, **419**, 375–377.

Laskar, J., Correia, A.C.M., Gastineau, M., *et al.* (2004). Long term evolution and chaotic diffusion of the insolation quantities of Mars. *Icarus*, **170**, 343–364.

Lebonnois, S., Quémerais, E., Montmessin, F., *et al.* (2006). Vertical distribution of ozone on Mars as measured by SPICAM/Mars Express using stellar occultations. *Journal of Geophysical Research*, **111**, E09S05, doi: 10.1029/2005JE002643.

Lee, D.-C. and Halliday, A.N. (1997). Core formation on Mars and differentiated asteroids. *Nature*, **388**, 854–857.

Lemoine, F.G., Smith, D.E., Rowlands, D.D., *et al.* (2001). An improved solution of the gravity field of Mars (GMM-2B) from Mars Global Surveyor. *Journal of Geophysical Research*, **106**, 23359–23376.

Lenardic, A., Nimmo, F., and Moresi, L. (2004). Growth of the hemispheric dichotomy and the cessation of plate tectonics on Mars. *Journal of Geophysical Research*, **109**, E02003, doi: 10.1029/2003JE002172.

Leovy, C. (2001). Weather and climate on Mars. *Nature*, **412**, 245–249.

Leshin, L.A. (2000). Insights into martian water reservoirs from analyses of martian meteorite QUE94201. *Geophysical Research Letters*, **27**, 2017–2020.

Lewis, K.W. and Aharonson, O. (2006). Stratigraphic analysis of the distributary fan in Eberswalde crater using stereo imagery. *Journal of Geophysical Research*, **111**, E06001, doi: 10.1029/2005JE002558.

Lewis, S.R., Collins, M., and Read, P.L. (1997). Data assimilation with a martian atmospheric GCM: an example using thermal data. *Advances in Space Research*, **19**, 1267–1270.

Lillis, R.J., Mitchell, D.L., Lin, R.P., Connerney, J.E.P., and Acuña, M.H. (2004). Mapping crustal magnetic fields at Mars using electron reflectometry. *Geophysical Research Letters*, **31**, L15702, doi: 10.1029/2004GL020189.

Lissauer, J.J., Dones, L., and Ohtsuki, K. (2000). Origin and evolution of terrestrial planet rotation. In *Origin of the Earth and Moon,* ed. R.M. Canup and K. Righter. Tucson, AZ: University of Arizona Press, pp.101–112.

Lodders, K. (1998). A survey of shergottite, nakhlite, and Chassigny meteorites whole-rock compositions. *Meteoritics and Planetary Science*, Suppl., **33**, A183–A190.

Lognonné, P. (2005). Planetary seismology. *Annual Reviews of Earth and Planetary Science*, **33**, 571–604.

Longhi, J. and Pan, V. (1989). The parent magmas of the SNC meteorites. In *Proceedings of the 19th Lunar and Planetary Science Conference*. Cambridge, UK: Cambridge University Press, pp.451–464.

Longhi, J., Knittle, E., Holloway, J.R., and Wänke, H. (1992). The bulk composition, mineralogy, and internal structure of Mars. In *Mars*, ed. H.H. Kieffer, B.M. Jakosky, C.W. Snyder, and M.S. Matthews. Tucson, AZ: University of Arizona Press, pp. 184–208.

Lucchitta, B.K. (1984). Ice and debris in the fretted terrain, Mars. *Journal of Geophysical Research*, **89**, B409–B418.

Lucchitta, B.K. (2001). Antarctic ice streams and outflow channels on Mars. *Geophysical Research Letters*, **28**, 403–406.

Lucchitta, B.K., McEwen, A.S., Clow, G.D., *et al*. (1992). The canyon system of Mars. In *Mars*, ed. H.H. Kieffer, B.M. Jakosky, C.W. Snyder, and M.S. Matthews. Tucson, AZ: University of Arizona Press, pp.453–492.

Lunine, J.I., Chambers, J., Morbidelli, A., and Leshin, L.A. (2003). The origin of water on Mars. *Icarus*, **165**, 1–8.

Lyons, J.R., Manning, C., and Nimmo, F. (2005). Formation of methane on Mars by fluid–rock interaction in the crust. *Geophysical Research Letters*, **32**, L13201, doi: 10.1029/2004GL022161.

Madsen, M.B., Hviid, S.F., Gunnlaugsson, H.P., *et al*. (1999). The magnetic properties experiments on Mars Pathfinder. *Journal of Geophysical Research*, **104**, 8761–8779.

Malin, M.C. and Carr, M.H. (1999). Groundwater formation of martian valleys. *Nature*, **397**, 589–591.

Malin, M.C. and Edgett, K.S. (1999). Oceans or seas in the martian northern lowlands: high resolution imaging tests of proposed coastlines. *Geophysical Research Letters*, **26**, 3049–3052.

Malin, M.C. and Edgett, K.S. (2000a). Sedimentary rocks of early Mars. *Science*, **290**, 1927–1937.

Malin, M.C. and Edgett, K.S. (2000b). Evidence for recent groundwater seepage and surface runoff on Mars. *Science*, **288**, 2330–2335.

Malin, M.C. and Edgett, K.S. (2001). Mars Global Surveyor Mars Orbiter Camera: interplanetary cruise through primary mission. *Journal of Geophysical Research*, **106**, 23429–23570.

Malin, M.C. and Edgett, K.S. (2003). Evidence for persistent flow and aqueous sedimentation on early Mars. *Science*, **302**, 1931–1934.

Malin, M.C., Edgett, K.S., Poslolova, L.V., McColley, S.M., and Noe Dobrea, E.Z. (2006). Present-day impact cratering rate and contemporary gully activity on Mars. *Science*, **314**, 1573–1577.

Mancinelli, R.L., Fahlen, T.F., Landheim, R., and Klovstad, M.R. (2004). Brines and evaporates: analogs for martian life. *Advances in Space Research*, **33**, 1244–1246.

Mangold, N. (2005). High latitude patterned grounds on Mars: classification, distribution and climatic control. *Icarus*, **174**, 336–359.

Mangold, N., Quantin, C., Ansan, V., Delacourt, C., and Allemand, P. (2004a). Evidence for precipitation on Mars from dendritic valleys on the Valles Marineris area. *Science*, **305**, 78–81.

Mangold, N., Maurice, S., Feldman, W.C., Costard, F., and Forget, F. (2004b). Spatial relationships between patterned ground and ground ice detected by the Neutron Spectrometer on Mars. *Journal of Geophysical Research*, **109**, E08001, doi: 10.1029/2004JE002235.

Mangus, S. and Larsen, W. (2004). *Lunar Receiving Laboratory Project History*, NASA/CR-2004-208938. Washington, DC: NASA.

Mantas, G.P. and Hanson, W.B. (1979). Photoelectron fluxes in the martian ionosphere. *Journal of Geophysical Research*, **84**, 369–385.

Mathew, K.J. and Marti, K. (2001). Early evolution of martian volatiles: nitrogen and noble gas components in ALH84001 and Chassigny. *Journal of Geophysical Research*, **106**, 1401–1422.

Max, M.D. and Clifford, S.M. (2000). The state, potential distribution, and biological implications of methane in the martian crust. *Journal of Geophysical Research*, **105**, 4165–4171.

Max, M.D. and Clifford, S.M. (2001). Initiation of martian outflow channels: related to the dissociation of gas hydrate? *Geophysical Research Letters*, **28**, 1787–1790.

McEwen, A.S. and Bierhaus, E.B. (2006). The importance of secondary cratering to age constraints on planetary surfaces. *Annual Reviews of Earth and Planetary Science*, **34**, 535–567.

McEwen, A.S., Malin, M.C., Carr, M.H., and Hartmann, W.K. (1999). Voluminous volcanism on early Mars revealed in Valles Marineris. *Nature*, **397**, 584–586.

McEwen, A.S., Preblich, B.S., Turtle, E.P., *et al.* (2005). The rayed crater Zunil and interpretations of small impact craters on Mars. *Icarus*, **176**, 351–381.

McGill, G.E. (2000). Crustal history of north central Arabia Terra, Mars. *Journal of Geophysical Research*, **105**, 6945–6959.

McGill, G.E. and Hills, L.S. (1992). Origin of giant martian polygons. *Journal of Geophysical Research*, **97**, 2633–2647.

McGovern, P.J., Solomon, S.C., Head, J.W., *et al.* (2001). Extension and uplift at Alba Patera, Mars: insights from MOLA observations and loading models. *Journal of Geophysical Research*, **106**, 23769–23809.

McGovern, P.J., Solomon, S.C., Smith, D.E., *et al.* (2002). Localized gravity/topography admittance and correlation spectra on Mars: implications for regional and global evolution. *Journal of Geophysical Research*, **107**, 5136, doi: 10.1029/2002JE001854.

McGovern, P.J., Solomon, S.C., Smith, D.E., *et al.* (2004a). Correction to "Localized gravity/topography admittance and correlation spectra on Mars: implications for regional and global evolution." *Journal of Geophysical Research*, **109**, E07007, doi: 10.1029/2004JE00286.

McGovern, P.J., Smith, J.R., Morgan, J.K., and Bulmer, M.H. (2004b). Olympus Mons aureole deposits: new evidence for a flank failure origin. *Journal of Geophysical Research*, **109**, E08008, doi: 10.1029/2004JE002258.

McKay, D.S., Gibson, E.K., Thomas-Keprta, K.L., *et al.* (1996). Search for past life on Mars: possible relic biogenic activity in martian meteorite ALH84001. *Science*, **273**, 924–930.

McSween, H.Y. (2002). The rocks of Mars, from far and near. *Meteoritics and Planetary Science*, **37**, 7–25.

McSween, H.Y. and Harvey, R.P. (1993). Outgassed water on Mars: constraints from melt inclusions in SNC meteorites. *Science*, **259**, 1890–1892.

McSween, H.Y., Murchie, S.L., Crisp, J.A., *et al.* (1999). Chemical, multispectral, and textural constraints on the composition and origin of rocks at the Mars Pathfinder landing site. *Journal of Geophysical Research*, **104**, 8679–8715.

McSween, H.Y., Grove, T.L., Lentz, R.C.F., *et al.* (2001). Geochemical evidence for magmatic water within Mars from pyroxenes in the Shergotty meteorite. *Nature*, **409**, 487–490.

McSween, H.Y., Grove, T.L., and Wyatt, M.B. (2003). Constraints on the composition and petrogenesis of the martian crust. *Journal of Geophysical Research*, **108**, 5135, doi: 10.1029/2003JE002175.

McSween, H.Y., Arvidson, R.E., Bell, J.F., *et al.* (2004). Basaltic rocks analyzed by the Spirit rover in Gusev Crater. *Science*, **305**, 842–845.

McSween, H.Y., Wyatt, M.B., Gellert, R., *et al.* (2006). Characterization and petrologic interpretation of olivine-rich basalts at Gusev Crater, Mars. *Journal of Geophysical Research*, **111**, E02S10, doi: 10.1029/2005JE002477.

Mège, D. and Masson, P. (1996). Amounts of crustal stretching in Valles Marineris, Mars. *Planetary and Space Science*, **44**, 749–782.

Meier, R., Owen, T.C., Matthews, H.E., *et al.* (1998). A determination of the HDO/$H_2O$ ratio in Comet C/1995 O1 (Hale-Bopp). *Science*, **279**, 842–844.

Mellon, M.T. and Jakosky, B.M. (1993). Geographic variations in the thermal and diffusive stability of ground ice on Mars. *Journal of Geophysical Research*, **98**, 3345–3364.

Mellon, M.T. and Jakosky, B.M. (1995). The distribution and behavior of martian ground ice during past and present epochs. *Journal of Geophysical Research*, **100**, 11781–11799.

Mellon, M.T., Jakosky, B.M., and Postawko, S.E. (1997). The persistence of equatorial ground ice on Mars. *Journal of Geophysical Research*, **102**, 19357–19369.

Mellon, M.T., Jakosky, B.M., Kieffer, H.H., and Christensen, P.R. (2000). High-resolution thermal inertia mapping from the Mars Global Surveyor Thermal Emission Spectrometer. *Icarus*, **148**, 437–455.

Mellon, M.T., Feldman, W.C., and Prettyman, T.H. (2004). The presence and stability of ground ice in the southern hemisphere of Mars. *Icarus*, **169**, 324–340.

Melosh, H.J. (1984). Impact ejection, spallation, and the origin of meteorites. *Icarus*, **59**, 234–260.

Melosh, H.J. (1989). *Impact Cratering: A Geologic Process*. New York: Oxford University Press.

Melosh, H.J. and Vickery, A.M. (1989). Impact erosion of the primordial atmosphere of Mars. *Nature*, **338**, 487–489.

Milkovich, S.M., Head, J.W., and Pratt, S. (2002). Meltback of Hesperian-aged ice-rich deposits near the south pole of Mars: evidence for drainage channels and lakes. *Journal of Geophysical Research*, **107**, 5043, doi: 10.1029/2001JE001802.

Ming, D.W., Mittlefehldt, D.W., Morris, R.V., *et al.* (2006). Geochemical and mineralogical indicators for aqueous processes in Columbia Hills of Gusev crater, Mars. *Journal of Geophysical Research*, **111**, E02S12, doi: 10.1029/2005JE002560.

Mischna, M.A., Richardson, M.I., Wilson, R.J., and McCleese, D.J. (2003). On the orbital forcing of martian water and $CO_2$ cycles: a general circulation model study with simplified volatile schemes. *Journal of Geophysical Research*, **108**, 5062, doi: 10.1029/2003JE002051.

Mitrofanov, I., Anfimov, D., Kozyrev, A., *et al.* (2002). Maps of subsurface hydrogen from the high energy neutron detector, Mars Odyssey. *Science*, **297**, 78–81.

Mittlefehldt, D.W. (1994). ALH84001, a cumulate orthopyroxenite member of the martian meteorite clan. *Meteoritics*, **29**, 214–221.

Montési, L.G.J. and Zuber, M.T. (2003). Clues to the lithospheric structure of Mars from wrinkle ridge sets and localization instability. *Journal of Geophysical Research*, **108**, 5048, doi: 10.1029/2002JE001974.

Moore, H.J. and Jakosky, B.M. (1989). Viking landing sites, remote-sensing observations, and physical properties of martian surface materials. *Icarus*, **81**, 164–184.

Moore, H.J., Bickler, D.B., Crisp, J.A., *et al.* (1999). Soil-like deposits observed by Sojourner, the Pathfinder rover. *Journal of Geophysical Research*, **104**, 8729–8746.

Moore, J.M. and Howard, A.D. (2005). Large alluvial fans on Mars. *Journal of Geophysical Research*, **110**, E04005, doi: 10.1029/2004JE002352.

Morbidelli, A., Chambers, J., Lunine, J.L., *et al.* (2000). Source regions and time scales for the delivery of water to Earth. *Meteoritics*, **35**, 1309–1320.

Morris, R.V., Golden, D.C., Bell, J.F., *et al.* (2000). Mineralogy, composition, and alteration of Mars Pathfinder rocks and soils: evidence from multispectral, elemental, and magnetic data on terrestrial analogue, SNC meteorite, and Pathfinder samples. *Journal of Geophysical Research*, **105**, 1757–1817.

Morris, R.V., Klingelhöfer, G., Bernhardt, B., *et al.* (2004). Mineralogy at Gusev Crater from the Mössbauer spectrometer on the Spirit rover. *Science*, **305**, 833–836.

Morris, R.V., Ming, D.W., Graff, T.G., *et al.* (2005). Hematite spherules in basaltic tephra altered under aqueous, acid-sulfate conditions on Mauna Kea volcano, Hawaii: possible clues for the occurrence of hematite-rich spherules in the Burns formation at Meridiani Planum, Mars. *Earth and Planetary Science Letters*, **240**, 168–178.

Morris, R.V., Klingelhöfer, G., Schröder, C., *et al.* (2006). Mössbauer mineralogy of rock, soil, and dust at Gusev Crater, Mars: Spirit's journey through weakly altered olivine basalt on the plains and pervasively altered basalt in the Columbia Hills. *Journal of Geophysical Research*, **111**, E02S13, doi: 10.1029/2005JE002584.

Mouginis-Mark, P.J. and Christensen, P.R. (2005). New observations of volcanic features on Mars from the THEMIS instrument. *Journal of Geophysical Research*, **110**, E08007, doi: 10.1029/2005JE002421.

Mouginis-Mark, P. and Yoshioka, M.T. (1998). The long lava flows of Elysium Planitia, Mars. *Journal of Geophysical Research*, **103**, 19389–19400.

Mouginis-Mark, P.J., Wilson, L., and Head, J.W. (1982). Explosive volcanism on Hecates Tholus, Mars: investigation of eruption conditions. *Journal of Geophysical Research*, **87**, 9890–9904.

Mouginis-Mark, P.J., Wilson, L., and Zimbelman, J.R. (1988). Polygenic eruptions on Alba Patera, Mars: evidence of channel erosion on pyroclastic flows. *Bulletin of Volcanology*, **50**, 361–379.

Mouginis-Mark, P.J., McCoy, T.J., Taylor, G.J., and Keil, K. (1992a). Martian parent craters for the SNC meteorites. *Journal of Geophysical Research*, **97**, 10213–10225.

Mouginis-Mark, P.J., Wilson, L., and Zuber, M.T. (1992b). The physical volcanology of Mars. In *Mars*, ed. H.H. Kieffer, B.M. Jakosky, C.W. Snyder, and M.S. Matthews. Tucson, AZ: University of Arizona Press, pp. 424–452.

Muhleman, D.O., Grossman, A.W., and Butler, B.J. (1995). Radar investigation of Mars, Mercury, and Titan. *Annual Reviews of Earth and Planetary Science*, **23**, 337–374.

Murchie, S. and Erard, S. (1996). Spectral properties and heterogeneity of Phobos from measurements by Phobos 2. *Icarus*, **123**, 63–86.

Murphy, J.R., Leovy, C.B., and Tillman, J.E. (1990). Observations of martian surface winds at the Viking Lander 1 site. *Journal of Geophysical Research*, **95**, 14555–14576.

Murphy, J.R., Pollack, J.B., Haberle, R.M., *et al.*, (1995). 3-dimensional numerical simulations of martian global dust storms. *Journal of Geophysical Research*, **100**, 26357–26376.

Murray, B., Koutnik, M., Byrne, S., *et al.* (2001). Preliminary geological assessment of the northern edge of Ultimi Lobe, Mars south polar layered deposits. *Icarus*, **154**, 80–97.

Murray, J.B., Muller, J.-P., Neukum, G., *et al.* (2005). Evidence from the Mars Express High Resolution Stereo Camera for a frozen sea close to Mars' equator. *Nature*, **434**, 352–356.

Musselwhite, D.S., Swindle, T.D., and Lunine, J.I. (2001). Liquid $CO_2$ breakout and the formation of recent small gullies on Mars. *Geophysical Research Letters*, **28**, 1283–1285.

Mustard, J.F. and Cooper, C.D. (2005). Joint analysis of ISM and TES spectra: the utility of multiple wavelength regimes for martian surface studies. *Journal of Geophysical Research*, **110**, E05012, doi: 10.1029/2004JE002355.

Mustard, J.F., Cooper, C.D., and Rifkin, M.K. (2001). Evidence for recent climate change on Mars from the identification of youthful near-surface ground ice. *Nature*, **412**, 411–414.

Mustard, J.F., Poulet, F., Gendrin, A., *et al.* (2005). Olivine and pyroxene diversity in the crust of Mars. *Science*, **307**, 1594–1597.

Mutch, T.A., Arvidson, R.E., Head, J.W., Jones, K.L., and Saunders, R.S. (1976). *The Geology of Mars*. Princeton, NJ: Princeton University Press.

Neukum, G., Jaumann, R., Hoffmann, H., *et al.* (2004). Recent and episodic volcanic and glacial activity on Mars revealed by the High Resolution Stereo Camera. *Nature*, **432**, 971–979.

Neumann, G.A., Zuber, M.T., Wieczorek, M.A., *et al.* (2004). Crustal structure of Mars from gravity and topography. *Journal of Geophysical Research*, **109**, E08002, doi: 10.1029/2004JE002262.

Newman, C.E., Lewis, S.R., Read, P.L., and Forget, F. (2002a). Modeling the martian dust cycle. 1. Representations of dust transport processes. *Journal of Geophysical Research*, **107**, 5123, doi: 10.1029/2002JE001910.

Newman, C.E., Lewis, S.R., Read, P.L., and Forget, F. (2002b). Modeling the martian dust cycle. 2. Multiannual radiatively active dust transport simulations. *Journal of Geophysical Research*, **107**, 5124, doi: 10.1029/2002JE001920.

Newman, M.J. and Rood, R.T. (1977). Implications of solar evolution for the Earth's early atmosphere. *Science*, **198**, 1035–1037.

Newsom, H.E., Hagerty, J.J., and Thorsos, I.E. (2001). Location and sampling of aqueous and hydrothermal deposits in martian impact craters. *Astrobiology*, **1**, 71–88.

Nimmo, F. (2000). Dike intrusion as a possible cause of linear martian magnetic anomalies. *Geology*, **28**, 391–394.

Nimmo, F. and Stevenson, D.J. (2000). Influence of early plate tectonics on the thermal evolution and magnetic field of Mars. *Journal of Geophysical Research*, **105**, 11969–11980.

Noe Dobrea, E.Z., Bell, J.F., Wolff, M.J., and Gordon, K.D. (2003). $H_2O$- and OH-bearing minerals in the martian regolith: analysis of 1997 observations from HST/NICMOS. *Icarus*, **166**, 1–20.

Norman, M.D. (1999). The composition and thickness of the crust of Mars estimated from REE and Nd isotopic compositions of martian meteorites. *Meteoritics and Planetary Science*, **34**, 439–449.

Norman, M.D. (2002). Thickness and composition of the martian crust revisited: implications of an ultradepleted mantle with a Nd isotopic composition like that of QUE94201. *Lunar and Planetary Science XXXIII*, Abstract #1157 (CD-ROM). Houston, TX: Lunar and Planetary Institute.

Novak, R.E., Mumma, M.J., DiSanti, M.A., Dello Russo, N., and Magee-Sauer, K. (2002). Mapping of ozone and water in the atmosphere of Mars near the 1997 aphelion. *Icarus*, **158**, 14–23.

Nye, J.F., Durham, W.B., Schenk, P.M., and Moore, J.M. (2000). The instability of a south polar cap on Mars composed of carbon dioxide. *Icarus*, **144**, 449–455.

Nyquist, L.E., Bogard, D.D., Shih, C.-Y., *et al.* (2001). Ages and geologic histories of martian meteorites. *Space Science Reviews*, **96**, 105–164.

Owen, T. and Bar-Nun, A. (1995). Comets, impacts, and atmospheres. *Icarus*, **116**, 215–226.

Owen, T., Maillard, J.P., deBergh, C., and Lutz, B.L. (1988). Deuterium on Mars: the abundance of HDO and the value of D/H. *Science*, **240**, 1767–1770.

Paige, D.A. (1992). The thermal stability of near-surface ground ice on Mars. *Nature*, **356**, 43–45.

Paige, D.A. and Keegan, K.D. (1994). Thermal and albedo mapping of the polar regions of Mars using Viking thermal mapper observations. 2. South polar region. *Journal of Geophysical Research*, **99**, 25993–26013.

Paige, D.A., Bachman, J.E., and Keegan, K.D. (1994). Thermal and albedo mapping of the polar regions of Mars using Viking thermal mapper observations. 1. North polar region. *Journal of Geophysical Research*, **99**, 25959–25991.

Parker, T.J., Saunders, R.S., and Schneeberger, D.M. (1989). Transitional morphology in west Deuteronilus Mensae, Mars: implications for modification of the lowland/upland boundary. *Icarus*, **82**, 111–145.

Parker, T.J., Gorsline, D.S., Saunders, R.S., Pieri, D.C., and Schneeberger, D.M. (1993). Coastal geomorphology of the martian northern plains. *Journal of Geophysical Research*, **98**, 11061–11078.

Pathare, A.V., Paige, D.A., and Turtle, E. (2005). Viscous relaxation of craters within the martian south polar layered deposits. *Icarus*, **174**, 396–418.

Pelkey, S.M., Jakosky, B.M., and Mellon, M.T. (2001). Thermal inertia of crater-related wind streaks on Mars. *Journal of Geophysical Research*, **106**, 23909–23920.

Pepin, R.O. (1991). On the origin and early evolution of terrestrial planet atmospheres and meteoritic volatiles. *Icarus*, **92**, 2–79.

Perrier, S., Bertaux, J.L., Lefèvre, F., *et al.* (2006). Global distribution of total ozone on Mars from SPICAM/MEX UV measurements. *Journal of Geophysical Research*, **111**, E09S06, doi: 10.1029/2006JE002681.

Perron, J.T., Dietrich, W.E., Howard, A.D., McKean, J.A., and Pettinga, J.R. (2003). Ice-driven creep on martian debris slopes. *Geophysical Research Letters*, **30**, 1747, doi: 10.1029/2003GL017603.

Petit, J.-M., Morbidelli, A., and Chambers, J. (2001). The primordial excitation and clearing of the asteroid belt. *Icarus*, **153**, 338–347.

Phillips, R.J. (1991). Expected rate of Marsquakes. In *Scientific Rationale and Requirements for a Global Seismic Network on Mars*, Technical Report 91–02. Houston, TX: Lunar and Planetary Institute, pp.35–38.

Phillips, R.J., Zuber, M.T., Solomon, S.C., *et al.* (2001). Ancient geodynamics and global-scale hydrology on Mars. *Science*, **291**, 2587–2591.

Picardi, G., Plaut, J.J., Biccari, D., *et al.* (2005). Radar soundings of the subsurface of Mars. *Science*, **310**, 1925–1928.

Pierazzo, E., Artemieva, N.A., and Ivanov, B.A. (2005). Starting conditions for hydrothermal systems underneath martian craters: Hydrocode modeling. In *Large Meteorite Impacts III*, Special Paper No. 384, ed. T. Kenkmann, F. Hörz, and A. Deutsch. Boulder, CO: Geological Society of America, pp.443–457.

Plescia, J.B. (1990). Recent flood lavas in the Elysium region of Mars. *Icarus*, **88**, 465–490.

Plescia, J.B. (1994). Geology of the small Tharsis volcanoes: Jovis Tholus, Ulysses Patera, Biblis Patera, Mars. *Icarus*, **111**, 246–269.

Plescia, J.B. (2000). Geology of the Uranius group volcanic constructs: Uranius Patera, Ceranunius Tholus, and Uranius Tholus. *Icarus*, **143**, 378–396.

Plescia, J.B. (2003). Cerberus Fossae, Elysium, Mars: a source for lava and water. *Icarus*, **164**, 79–95.

Pollack, J.B., Burns, J.A., and Tauber, M.E. (1979). Gas drag in primordial circumplanetary envelopes: a mechanism for satellite capture. *Icarus*, **37**, 587–611.

Pollack, J.B., Kasting, J.F., Richardson, S.M., and Poliakoff, K. (1987). The case for a warm wet climate on early Mars. *Icarus*, **71**, 203–224.

Pollack, J.B., Haberle, R.M., Schaeffer, J., and Lee, H. (1990). Simulations of the general circulation of the martian atmosphere. 1. Polar processes. *Journal of Geophysical Research*, **95**, 1447–1473.

Pollack, J.B., Haberle, R.M., Murphy, J.R., Schaeffer, H., and Lee, H. (1993). Simulations of the general circulation of the martian atmosphere. 2. Seasonal pressure variations. *Journal of Geophysical Research*, **98**, 3149–3181.

Poulet, F., Bibring, J.-P., Mustard, J.F., *et al.* (2005). Phyllosilicates on Mars and implications for early martian climate. *Nature*, **438**, 623–627.

Pruis, M.J. and Tanaka, K.L. (1995). The martian northern plains did not result from plate tectonics. In *Lunar and Planetary Science XXVI*. Houston, TX: Lunar and Planetary Institute, pp.1147–1148.

Putzig, N.E., Mellon, M.T., Kretke, K.A., and Arvidson, R.E. (2005). Global thermal inertia and surface properties of Mars from the MGS mapping mission. *Icarus*, **173**, 325–341.

Quantin, C., Allemand, P., Mangold, N., and Delacourt, C. (2004). Ages of Valles Marineris (Mars) landslides and implications for canyon history. *Icarus*, **172**, 555–572.

Race, M.S. (1996). Planetary protection, legal ambiguity and the decision making process for Mars sample return. *Advances in Space Research*, **18**, 345–350.

Rafkin, S.C.R., Haberle, R.M., and Michaels, T.I. (2001). The Mars Regional Atmospheric Modeling System: model description and selected simulations. *Icarus*, **151**, 228–256.

Read, P.L. and Lewis, S.R. (2004). *The Martian Climate Revisited*. Chichester, UK: Praxis Publishing.

Reid, I.N., Sparks, W.B., Lubow, S., *et al.* (2006). Terrestrial models for extraterrestrial life: methanogens and halophiles at martian temperatures. *International Journal of Astrobiology*, **5**, 89–97.

Richardson, M.I. and Mischna, M.A. (2005). Long-term evolution of transient liquid water on Mars. *Journal of Geophysical Research*, **110**, E03003, doi: 10.1029/2004JE002367.

Richardson, M.I. and Wilson, R.J. (2002). Investigation of the nature and stability of the martian seasonal water cycle with a general circulation model. *Journal of Geophysical Research*, **107**, 5031, doi: 10.1029/2001JE001536.

Richardson, M.I., Wilson, R.J., and Rodin, V. (2002). Water ice clouds in the martian atmosphere: general circulation model experiments with a simple cloud scheme. *Journal of Geophysical Research*, **107**, 5064, doi: 10.1029/2001JE001804.

Rieder, R., Wänke, H., Economou, T., and Turkevich, A. (1997a). Determination of the chemical composition of martian soil and rocks: the alpha proton X-ray spectrometer. *Journal of Geophysical Research*, **102**, 4027–4044.

Rieder, R., Economou, T., Wänke, H., *et al.* (1997b). The chemical composition of martian soil and rocks returned by the mobile alpha proton X-ray spectrometer: preliminary results from the X-ray mode. *Science*, **278**, 1771–1774.

Rieder, R., Gellert, R., Brückner, J., *et al.* (2003). The new Athena alpha particle X-ray spectrometer for the Mars Exploration Rovers. *Journal of Geophysical Research*, **108**, 8066, doi: 10.1029/2003JE002150.

Rieder, R., Gellert, R., Anderson, R.C., *et al.* (2004). Chemistry of rocks and soils at Meridiani Planum from the alpha particle X-ray spectrometer. *Science*, **306**, 1746–1749.

Righter, K., Hervig, R.J., and Kring, D.A. (1998). Accretion and core formation on Mars: molybdenum contents of melt inclusion glasses in three SNC meteorites. *Geochimica et Cosmochimica Acta*, **62**, 2167–2177.

Rivkin, A.S., Binzel, R.P., Howell, E.S., Bus, S.J., and Grier, J.A. (2003). Spectroscopy and photometry of Mars Trojans. *Icarus*, **165**, 349–354.

Robert, F. (2001). The origin of water on Earth. *Science*, **293**, 1056–1058.

Robert, F., Gautier, D., and Dubrulle, B. (2000). The solar system D/H ratio: observations and theories. *Space Science Reviews*, **92**, 201–224.

Robinson, M.S., Mouginis-Mark, P.J., Zimbelman, J.R., *et al.* (1993). Chronology, eruption duration, and atmospheric contribution of the martian volcano Apollinaris Patera. *Icarus*, **104**, 301–323.

Rochette, P., Fillion, G., Ballou, R., *et al.* (2003). High pressure magnetic transition in pyrrhotite and impact demagnetization on Mars. *Geophysical Research Letters*, **30**, 1683, doi: 10.1029/2003GL017359.

Roddy, D.J., Pepin, R.O., and Merrill, R.B. (1977). *Impact and Explosion Cratering: Planetary and Terrestrial Implications*. New York: Pergamon Press.

Romanek, C.S., Grady, M.M., Wright, I.P., *et al.* (1994). Record of fluid–rock interactions on Mars from the meteorite ALH84001. *Nature*, **37**, 655–657.

Rubie, D.C., Melosh, H.J., Reid, J.E., Liebske, C., and Righter, K. (2003). Mechanisms of metal–silicate equilibrium in the terrestrial magma ocean. *Earth and Planetary Science Letters*, **205**, 239–255.

Rummel, J.D. and Meyer, M.A. (1996). A consensus approach to planetary protection requirements: recommendations for Mars Lander Missions. *Advances in Space Research*, **18**, 317–321.

Rummel, J.D., Stabekis, P.D., DeVincenzi, D.L., and Barengoltz, J.B. (2002). COSPAR's planetary protection policy: a consolidated draft. *Advances in Space Research*, **30**, 1567–1571.

Ryan, S., Dlugokencky, E.J., Tans, P.P., and Trudeau, M.E. (2006). Mauna Loa volcano is not a methane source: implications for Mars. *Geophysical Research Letters*, **33**, L12301, doi: 10.1029/2006GL026223.

Sagan, C. and Chyba, C. (1997). The early faint sun paradox: organic shielding of ultraviolet-labile greenhouse gases. *Science*, **276**, 1217–1221.

Schaber, G.G. (1982). Syrtis Major: a low-relief volcanic shield. *Journal of Geophysical Research*, **87**, 9852–9866.

Schneeberget, D.M. and Pieri, D.C. (1991). Geomorphology and stratigraphy of Alba Patera, Mars. *Journal of Geophysical Research*, **98**, 1907–1930.

Schofield, J.T., Barnes, J.R., Crisp, D., *et al.* (1997). The Mars Pathfinder Atmospheric Structure Investigation/Meteorology (ASI/MET) experiment. *Science*, **278**, 1752–1758.

Scholl, H., Marzari, F., and Tricarico, P. (2005). Dynamics of Mars Trojans. *Icarus*, **175**, 397–408.

Schorghofer, N. and Aharonson, O. (2005). Stability and exchange of subsurface ice on Mars. *Journal of Geophysical Research*, **110**, E05003, doi: 10.1029/2004JE002350.

Schubert, G., Russell, C.T., and Moore, W.B. (2000). Timing of the martian dynamo. *Nature*, **408**, 666–667.

Schuerger, A.C., Richards, J.T., Newcombe, D.A., and Venkateswaran, K. (2006). Rapid inactivation of seven *Bacillus* spp. under simulated Mars UV irradiation. *Icarus*, **181**, 52–62.

Schultz, P.H. (1992). Atmospheric effects on ejecta emplacement. *Journal of Geophysical Research*, **97**, 11623–11662.

Schultz, P.H. and Lutz-Garihan, A.B. (1982). Grazing impacts on Mars: a record on lost satellites. *Journal of Geophysical Research*, **87**, A84–A96.

Schultz, R.A. (1998). Multiple-process origin of Valles Marineris basins and troughs, Mars. *Planetary and Space Science*, **46**, 827–834.

Schultz, R.A. (2000). Localization of bedding plane slip and backthrust faults above blind thrust faults: keys to wrinkle ridge structure. *Journal of Geophysical Research*, **105**, 12035–12052.

Scott, D.H. and Tanaka, K.L. (1982). Ignimbrites of Amazonis Planitia region of Mars. *Journal of Geophysical Research*, **87**, 1179–1190.

Scott, E.R.D. (1999). Origin of carbonate–magnetite–sulfide assemblages in martian meteorite ALH84001. *Journal of Geophysical Research*, **104**, 3803–3813.

Scott, E.R.D. and Fuller, M. (2004). A possible source for the martian crustal magnetic field. *Earth and Planetary Science Letters*, **220**, 83–90.

Sears, D.W.G. and Kral, T.A. (1998). Martian "microfossils" in lunar meteorites? *Meteoritics and Planetary Science*, **33**, 791–794.

Segura, T.L., Toon, O.B., Colaprete, A., and Zahnle, K. (2002). Environmental effects of large impacts on Mars. *Science*, **298**, 1977–1980.

Seu, R., Biccari, D., Orosei, R., *et al.* (2004). SHARAD: the MRO 2005 shallow radar. *Planetary and Space Science*, **52**, 157–166.

Shih, C.-Y., Nyquist, L.E., Bogard, D.D., *et al.* (1982). Chronology and petrogenesis of young achondrites, Shergotty, Zagami, and ALH77005: late magmatism on a geologically active planet. *Geochimica et Cosmochimica Acta*, **46**, 2323–2344.

Shih, C.-Y., Nyquist, L.E., and Wiesmann, H. (1999). Samarium–neodymium and rubidium–strontium systematics of nakhlite Governador Valadares. *Meteoritics*, **34**, 647–655.

Shoemaker, E.M. (1963). Impact mechanics at Meteor Crater, Arizona. In *The Moon, Meteorites, and Comets*, ed. B.M. Middlehurst and G.P. Kuiper. Chicago, IL: University of Chicago Press, pp. 301–336.

Shoemaker, E.M. and Chao, E.C.T. (1962). New evidence for the impact origin of the Ries Basin, Bavaria, Germany. *Journal of Geophysical Research*, **66**, 3371–3378

Siebert, N.M. and Kargel, J.S. (2001). Small-scale martian polygonal terrain: implications for liquid surface water. *Geophysical Research Letters*, **28**, 899–902.

Simonelli, D.P., Wisz, M., Switala, A., *et al.* (1998). Photometric properties of Phobos surface materials from Viking images. *Icarus*, **131**, 52–77.

Simpson, R.A., Harmon, J.K., Zisk, S.H., Thompson, T.W., and Muhleman, D.O. (1992). Radar determinations of Mars surface properties. In *Mars*, ed. H.H. Kieffer, B.M. Jakosky, C.W. Snyder, and M.S. Matthews. Tucson, AZ: University of Arizona Press, pp. 652–685.

Singer, R.B. (1985). Spectroscopic observation of Mars. *Advances in Space Research*, **5**, 59–68.

Sizemore, H.G. and Mellon, M.T. (2006). Effects of soil heterogeneity on martian ground-ice stability and orbital estimates of ice table depth. *Icarus*, **185**, 358–369.

Sleep, N.H. (1994). Martian plate tectonics. *Journal of Geophysical Research*, **99**, 5639–5655.

Smith, D.E., Zuber, M.T., Solomon, S.C., *et al.* (1999). The global topography of Mars and implications for surface evolution. *Science*, **284**, 1495–1503.

Smith, D.E., Zuber, M.T., Frey, H.V., *et al.* (2001a). Mars Orbiter Laser Altimeter: experiment summary after the first year of global mapping of Mars. *Journal of Geophysical Research*, **106**, 23689–23722.

Smith, D.E., Zuber, M.T., and Neumann, G.A. (2001b). Seasonal variations of snow depth on Mars. *Science*, **294**, 2141–2146.

Smith, M.D., Pearl, J.C., Conrath, B.J., and Christensen, P.R. (2000). Mars Global Surveyor Thermal Emission Spectrometer (TES) observations of dust opacity during aerobraking and science phasing. *Journal of Geophysical Research*, **105**, 9539–9552.

Smith, P.H. and the Phoenix Science Team (2004). The Phoenix mission to Mars. *In Lunar and Planetary Science XXXV*, Abstract #2050. Houston, TX: Lunar and Planetary Institute.

Smith, P.H., Bell, J.F., Bridges, N.T., *et al.* (1997). Results from the Mars Pathfinder Camera. *Science*, **278**, 1758–1765.

Snyder Hale, A., Bass, D.S., and Tamppari, L.K. (2005). Monitoring the perennial martian northern polar cap with MGS MOC. *Icarus*, **174**, 502–512.

Soare, R.J., Burr, D.M., and Tseung, J.M.W.B. (2005). Possible pingos and a periglacial landscape in northwest Utopia Planitia. *Icarus*, **174**, 373–382.

Soderblom, L.A. (1992). The composition and mineralogy of the martian surface from spectroscopic observations: 0.3 μm to 50 μm. In *Mars*, ed. H.H. Kieffer, B.M. Jakosky, C.W. Snyder, and M.S. Matthews. Tucson, AZ: University of Arizona Press, pp. 557–593.

Soderblom, L.A., Anderson, R.C., Arvidson, R.E., *et al.* (2004). Soils of Eagle Crater and Meridiani Planum at the Opportunity rover landing site. *Science*, **306**, 1723–1726.

Soffen, G.A. (1977). The Viking Project. *Journal of Geophysical Research*, **82**, 3959–3970.

Sohl, F. and Spohn, T. (1997). The structure of Mars: implications from SNC meteorites. *Journal of Geophysical Research*, **102**, 1613–1635.

Solomatov, V.S. (2000). Fluid dynamics of a terrestrial magma ocean. In *Origin of the Earth and Moon*, ed. R.M. Canup and K. Righter. Tucson, AZ: University of Arizona Press, pp. 323–338.

Solomon, S.C. (1979). Formation, history, and energetics of cores in the terrestrial planets. *Earth and Planetary Science Letters*, **19**, 168–182.

Solomon, S.C. and Head, J.W. (1981). The importance of heterogenous lithospheric thickness and volcanic construction. *Journal of Geophysical Research*, **82**, 9755–9774.

Solomon, S.C., Aharonson, O., Aurnou, J.M., *et al.* (2005). New perspectives on ancient Mars. *Science*, **307**, 1214–1220.

Soukhovitskaya, V. and Manga, M. (2006). Martian landslides in Valles Marineris: wet or dry? *Icarus*, **180**, 348–352.

Space Studies Board (1977). *Post-Viking Biological Investigations of Mars*. Washington, DC: National Academy of Sciences.

Sprenke, K.F. and Baker, L.L. (2000). Magnetization, paleomagnetic poles, and polar wander on Mars. *Icarus*, **147**, 26–34.

Squyres, S.W. and Carr, M.H. (1986). Geomorphic evidence for the distribution of ground ice on Mars. *Science*, **231**, 249–252.

Squyres, S.W. and Kasting, J.F. (1994). Early Mars: how warm and how wet? *Science*, **265**, 744–749.

Squyres, S.W., Arvidson, R.E., Bell, J.F., *et al.* (2004a). The Spirit rover's Athena science investigation at Gusev Crater, Mars. *Science*, **305**, 794–799.

Squyres, S.W., Arvidson, R.E., Bell, J.F., *et al.* (2004b). The Opportunity rover's Athena science investigation at Meridiani Planum, Mars. *Science*, **306**, 1698–1703.

Squyres, S.W., Grotzinger, J.P., Arvidson, R.E., *et al.* (2004c). In situ evidence for an ancient aqueous environment at Meridiani Planum, Mars. *Science*, **306**, 1709–1714.

Squyres, S.W., Arvidson, R.E., Blaney, D.L., *et al.* (2006). Rocks of the Columbia Hills. *Journal of Geophysical Research*, **111**, E02S11, doi: 10.1029/2005JE002562.

Stevenson, D.J. (2001). Mars' core and magnetism. *Nature*, **412**, 214–219.

Stevenson, D.J. (2003). Planetary magnetic fields. *Earth and Planetary Science Letters*, **208**, 1–11.

Stewart, S.T. and Nimmo, F. (2002). Surface runoff features on Mars: testing the carbon dioxide formation hypothesis. *Journal of Geophysical Research*, **107**, 5069, doi: 10.1029/2000JE001465.

Stewart, S.T., O'Keefe, J.D., and Ahrens, T.J. (2001). The relationship between rampart crater morphologies and the amount of subsurface ice. In *Lunar and Planetary Science XXXII*, Abstract #2092. Houston, TX: Lunar and Planetary Institute.

Stöffler, D. and Ryder, G. (2001). Stratigraphy and isotope ages of lunar geologic units: chronological standard for the inner solar system. *Space Science Reviews*, **96**, 9–54.

Strom, R.G., Croft, S.K., and Barlow, N.G. (1992). The martian impact cratering record. In *Mars*, ed. H.H. Kieffer, B.M. Jakosky, C.W. Snyder, and M.S. Matthews. Tucson, AZ: University of Arizona Press, pp. 383–423.

Strom, R.G., Malhotra, R., Ito, T., Yoshida, F., and Kring, D.A. (2005). The origin of planetary impactors in the inner solar system. *Science*, **309**, 1847–1850.

Sullivan, R., Thomas, P., Veverka, J., Malin, M., and Edgett, K.S. (2001). Mass movement slope streaks imaged by the Mars Orbiter Camera. *Journal of Geophysical Research*, **106**, 23607–23633.

Sullivan, R., Bandfield, D., Bell, J.F., *et al.* (2005). Aeolian processes at the Mars Exploration Rover Meridiani Planum landing site. *Nature*, **436**, 58–61.

Swindle, T.D. and Jones, J.H. (1997). The xenon isotopic composition of the primordial martian atmosphere: contributions from solar and fission components. *Journal of Geophysical Research*, **102**, 1671–1678.

Tabuchnik, K.S. and Evans, N.W. (1999). Cartography for martian Trojans. *Astrophysical Journal*, **517**, L63–L66.

Tanaka, K.L. and Kolb, E.J. (2001). Geologic history of the polar regions of Mars based on Mars Global Surveyor Data. 1. Noachian and Hesperian periods. *Icarus*, **154**, 3–21.

Tanaka, K.L., Scott, D.H., and Greeley, R. (1992). Global stratigraphy. In *Mars*, ed. H.H. Kieffer, B.M. Jakosky, C.W. Snyder, and M.S. Matthews. Tucson, AZ: University of Arizona Press, pp. 345–382.

Taylor, G.J., Martel, L.M.V., and Boynton, W.V. (2006). Mapping Mars geochemically. In *Lunar and Planetary Science XXXVII*, Abstract #1981. Houston, TX: Lunar and Planetary Institute.

Tera, F., Papanastassiou, D.A., and Wasserburg, G.J. (1974). Isotopic evidence for a terminal lunar cataclysm. *Earth and Planetary Science Letters*, **22**, 1–21.

Terasaki, H., Frost, D.J., Rubie, D.C., and Langenhorst, F. (2005). The effect of oxygen and sulphur on the dihedral angle between Fe-O-S melt and silicate minerals at high pressure: implications for martian core formation. *Earth and Planetary Science Letters*, **232**, 379–392.

Thomas, P., Veverka, J., Bell, J., Lunine, J., and Cruikshank, D. (1992a). Satellites of Mars: geologic history. In *Mars*, ed. H.H. Kieffer, B.M. Jakosky, C.W. Snyder, and M.S. Matthews. Tucson, AZ: University of Arizona Press, pp.1257–1282.

Thomas, P., Squyres, S., Herkenhoff, K., Howard, A., and Murray, B. (1992b). Polar deposits of Mars. In *Mars*, ed. H.H. Kieffer, B.M. Jakosky, C.W. Snyder, and M.S. Matthews. Tucson, AZ: University of Arizona Press, pp.767–795.

Thomas, P.C., Veverka, J., Sullivan, R., *et al.* (2000a). Phobos: regolith and ejecta blocks investigated with Mars Orbiter Camera images. *Journal of Geophysical Research*, **105**, 15091–15106.

Thomas, P.C., Malin, M.C., Edgett, K.S., *et al.* (2000b). North–south geological differences between the residual polar caps on Mars. *Nature*, **404**, 161–164.

Thomas, P.C., Gierasch, P., Sullivan, R., *et al.* (2003). Mesoscale linear streaks on Mars: environments of dust entrainment. *Icarus*, **162**, 242–258.

Thomas, P.C., Malin, M.C., James, P.B., *et al.* (2005). South polar residual cap of Mars: features, stratigraphy, and changes. *Icarus*, **174**, 535–559.

Thomas-Keprta, K.L., Bazylinski, D.A., Kirschvink, J.L., *et al.* (2000). Elongated prismatic magnetite crystals in ALH84001 carbonate globules: potential martian magnetofossils. *Geochimica et Cosmochimica Acta*, **64**, 4049–4081.

Thomson, B.J. and Head, J.W. (2001). Utopia basin, Mars: characterization of topography and morphology and assessment of the origin and evolution of basin internal structures. *Journal of Geophysical Research*, **106**, 23209–23230.

Titus, T.N., Kieffer, H.H., Mullins, K.F., and Christensen, P.R. (2001). TES premapping data: slab ice and snow flurries in the martian north polar night. *Journal of Geophysical Research*, **106**, 23181–23196.

Titus, T.N., Kieffer, H.H., and Christensen, P.R. (2003). Exposed water ice discovered near the south pole of Mars. *Science*, **299**, 1048–1050.

Toigo, A.D. and Richardson, M.I. (2002). A mesoscale model for the martian atmosphere. *Journal of Geophysical Research*, **107**, 5049, doi: 10.1029/2001JE001489.

Toigo, A.D. and Richardson, M.I. (2003). Meteorology of proposed Mars Exploration Rover landing sites. *Journal of Geophysical Research*, **108**, 8092, doi: 10.1029/2003JE002064.

Toigo, A.D., Richardson, M.I., Ewald, S.P., and Gierasch, P.J. (2003). Numerical simulations of martian dust devils. *Journal of Geophysical Research*, **108**, 5047, doi: 10.1029/2002JE002002.

Toksöz, M.N. and Hsui, A.T. (1978). Thermal history and evolution of Mars. *Icarus*, **34**, 537–547.

Tonks, W.B. and Melosh, H.J. (1990). The physics of crystal settling and suspension in a turbulent magma ocean. In *Origin of the Earth*, ed. H.E. Newsom and J.H. Jones. New York: Oxford University Press, pp.151–174.

Tornabene, L.L., Moersch, J.E., McSween, H.Y., *et al.* (2006). Identification of large (2–10 km) rayed craters on Mars in THEMIS thermal infrared images: implications for possible martian meteorite source regions. *Journal of Geophysical Research*, **111**, E10006, doi: 10.1029/2005JE002600.

Touma, J. and Wisdom, J. (1993). The chaotic obliquity of Mars. *Science*, **259**, 1294–1296.

Treiman, A.H. (1995). S ≠ NC: multiple source areas for martian meteorites. *Journal of Geophysical Research*, **100**, 5329–5340.

Treiman, A.H., Barrett, R.A., and Gooding, J.L. (1993). Preterrestrial aqueous alteration of the Lafayette (SNC) meteorite. *Meteoritics*, **28**, 86–97.

Treiman, A.H., Fuks, K.H., and Murchie, S. (1995). Diagenetic layers in the upper walls of Valles Marineris, Mars: evidence for drastic climate change since the mid-Hesperian. *Journal of Geophysical Research*, **100**, 26339–26344.

Trofimov, V.I., Victorov, A., and Ivanov, M. (1996). Selection of sterilization methods for planetary return missions. *Advances in Space Research*, **18**, 333–337.

Turcotte, D.L. and Schubert, G. (2002). *Geodynamics*, 2nd edn. Cambridge, UK: Cambridge University Press.

Turtle, E.P. and Melosh, H.J. (1997). Stress and flexural modeling of the martian lithospheric response to Alba Patera. *Icarus*, **126**, 197–211.

Tyler, D., Barnes, J.R., and Haberle, R.M. (2002). Simulation of surface meteorology at the Pathfinder and VL1 sites using a Mars mesoscale model. *Journal of Geophysical Research*, **107**, 5018, doi: 10.1029/2001JE001618.

Vago, J.L., Gardini, B., Baglioni, P., *et al.* (2006). ExoMars: ESA's mission to search for signs of life on the Red Planet. In *Lunar and Planetary Science XXXVII*, Abstract #1871. Houston, TX: Lunar and Planetary Institute.

van Gasselt, S., Reiss, D., Thorpe, A.K., and Neukum, G. (2005). Seasonal variations of polygonal thermal contraction crack patterns in a south polar trough, Mars. *Journal of Geophysical Research*, **110**, E08002, doi: 10.1029/2004JE002385.

van Thienen, P., Vlaar, N.J., and van den Berg, A.P. (2004). Plate tectonics on the terrestrial planets. *Physics of the Earth and Planetary Interiors*, **142**, 61–74.

Varnes, E.S., Jakosky, B.M., and McCollom, T.M. (2003). Biological potential of martian hydrothermal systems. *Astrobiology*, **3**, 407–414.

Vasavada, A.R. and the MSL Science Team (2006). NASA's 2009 Mars Science Laboratory: an update. In *Lunar and Planetary Science XXXVII*, Abstract #1940. Houston, TX: Lunar and Planetary Institute.

Vasavada, A.R., Williams, J.-P., Paige, D.A., *et al.* (2000). Surface properties of Mars' polar layered deposits and polar landing sites. *Journal of Geophysical Research*, **105**, 6961–6969.

Wallace, P. and Carmichael, I.S.E. (1992). Sulfur in basaltic magmas. *Geochimica et Cosmochimica Acta*, **56**, 1863–1874.

Wang, H. and Ingersoll, A.P. (2002). Martian clouds observed by Mars Global Surveyor Mars Orbiter Camera. *Journal of Geophysical Research*, **107**, 5078, doi: 10.1029/2001JE001815.

Wang, H., Zurek, R.W., and Richardson, M.I. (2005). Relationship between frontal dust storms and transient eddy activity in the northern hemisphere of Mars as observed by Mars Global Surveyor. *Journal of Geophysical Research*, **110**, E07005, doi: 10.1029/2005JE002423.

Wänke, H. (1981). Constitution of terrestrial planets. *Philosophical Transactions of the Royal Society of London A*, **303**, 287–302.

Ward, A.W. (1979). Yardangs on Mars: evidence of recent wind erosion. *Journal of Geophysical Research*, **84**, 8147–8166.

Ward, W.R. (1992). Long-term orbital and spin dynamics of Mars. In *Mars*, ed. H.H. Kieffer, B.M. Jakosky, C.W. Snyder, and M.S. Matthews. Tucson, AZ: University of Arizona Press, pp. 298–320.

Ward, W.R. (2000). On planetesimal formation: the role of collective particle behavior. In *Origin of the Earth and Moon*, ed. R.M. Canup and K. Righter. Tucson, AZ: University of Arizona Press, pp. 75–84.

Warren, P.H. (1998). Petrologic evidence for low-temperature, possibly flood-evaporitic origin of carbonates in the ALH 84001 meteorite. *Journal of Geophysical Research*, **103**, 16759–16773.

Watson, L.L., Hutcheon, I.D., Epstein, S., and Stolper, E.M. (1994). Water on Mars: clues from deuterium/hydrogen and water contents of hydrous phases in SNC meteorites. *Science*, **265**, 86–90.

Watters, T.R. (2003). Thrust faults along the dichotomy boundary in the eastern hemisphere of Mars. *Journal of Geophysical Research*, **108**, 5054, doi: 10.1029/2002JE001934.

Watters, T.R. (2004). Elastic dislocation modeling of wrinkle ridges on Mars. *Icarus*, **171**, 284–294.

Watters, T.R., Leuschen, C.J., Plaut, J.J., *et al.* (2006). MARSIS radar sounder evidence of buried basins in the northern lowlands of Mars. *Nature*, **444**, 905–908.

Weidenschilling, S.J. and Cuzzi, J.N. (1993). Formation of planetesimals in the solar nebula. In *Protostars and Planets III*, ed. E.H. Levy and J.I. Lunine. Tucson, AZ: University of Arizona Press, pp. 1031–1060.

Wentworth, S.J., Gibson, E.K., Velbel, M.A., and McKay, D.S. (2005). Antarctic dry valleys and indigenous weathering in Mars meteorites: implications for water and life on Mars. *Icarus*, **174**, 383–395.

Werner, S.C., van Gasselt, S., and Neukum, G. (2003). Continual geological activity in Athabasca Valles, Mars. *Journal of Geophysical Research*, **108**, 8081, doi: 10.1029/2002JE002020.

Werner, S.C., Ivanov, B.A., and Neukum, G. (2006). Mars: secondary cratering – implications for age determination. In *Workshop on Surface Ages and Histories: Issues in Planetary Chronology*, Contribution No. 1320. Houston, TX: Lunar and Planetary Institute, pp. 55–56.

Wetherill, G.W. (1990). Formation of the Earth. *Annual Reviews of Earth and Planetary Science*, **18**, 205–256.

Wetherill, G.W. and Inaba, S. (2000). Planetary accumulation with a continuous supply of planetesimals. *Space Science Reviews*, **92**, 311–320.

Whelley, P.L. and Greeley, R. (2006). Latitudinal dependency in dust devil activity on Mars. *Journal of Geophysical Research*, **111**, E10003, doi: 10.1029/2006JE002677.

Whitmire, D.P., Doyle, L.R., Reynolds, R.T., and Matese, J.J. (1995). A slightly more massive young Sun as an explanation for warm temperatures on early Mars. *Journal of Geophysical Research*, **100**, 5457–5464.

Wieczorek, M.A. and Zuber, M.T. (2004). Thickness of the martian crust: improved constraints from geoid-to-topography ratios. *Journal of Geophysical Research*, **109**, E01009, doi: 10.1029/2003JE002153.

Wilhelms, D.E. and Squyres, S.W. (1984). The martian hemispheric dichotomy may be due to a giant impact. *Nature*, **309**, 138–140.

Williams, J.-P. and Nimmo, F. (2004). Thermal evolution of the martian core: implications for an early dynamo. *Geology*, **32**, 97–100.

Williams, J.-P., Paige, D.A., and Manning, C.E. (2003). Layering in the wall rock of Valles Marineris: intrusive and extrusive magmatism. *Geophysical Research Letters*, **30**, 1623, doi: 10.1029/2003GL017662.

Wilson, L. and Head, J.W. (1994). Mars: review and analysis of volcanic eruption theory and relationships to observed landforms. *Review of Geophysics*, **32**, 221–263.

Wilson, L. and Head, J.W. (2002). Tharsis-radial graben systems as the surface manifestation of plume-related dike intrusion complexes: models and implications. *Journal of Geophysical Research*, **107**, 5057, doi: 10.1029/2001JE001593.

Wilson, R.J. (1997). A general circulation model simulation of the martian polar warming. *Geophysical Research Letters*, **24**, 123–126.

Wilson, R.J., Banfield, D., Conrath, B.J., and Smith, M.D. (2002). Traveling waves in the northern hemisphere of Mars. *Geophysical Research Letters*, **29**, 1684, doi: 10.1029/2002GL014866.

Wise, D.U., Golombek, M.P., and McGill, G.E. (1979). Tectonic evolution of Mars. *Journal of Geophysical Research*, **84**, 7934–7939.

Wood, C.A. and Ashwal, L.D. (1981). SNC Meteorites: igneous rocks from Mars. In *Proceedings of the 12th Lunar and Planetary Science Conference*. New York: Pergamon Press, pp.1359–1375.

Wood, C.A., Head, J.W., and Cintala, M.J. (1978). Interior morphology of fresh martian craters: the effects of target characteristics. In *Proceedings of the 9th Lunar and Planetary Science Conference*. New York: Pergamon Press, pp.3691–3709.

Wyatt, M.B. and McSween, H.Y. (2002). Spectral evidence for weathered basalt as an alternative to andesite in the northern lowlands of Mars. *Nature*, **417**, 263–266.

Yen, A.S., Gellert, R., Schröder, C., *et al.* (2005). An integrated view of the chemistry and mineralogy of martian soils. *Nature*, **436**, 49–54.

Yoder, C.F., Konopliv, A.S., Yuan, D.N., Standish, E.M., and Folkner, W.M. (2003). Fluid core size of Mars from detection of the solar tide. *Science,* **300**, 299–303.

Youdin, A.N. and Shu, F.H. (2002). Planetesimal formation by gravitational instability. *Astrophysical Journal*, **580**, 494–505.

Yung, Y.L., Nair, H., and Gerstell, M.F. (1997). $CO_2$ greenhouse in the early martian atmosphere: $SO_2$ inhibits condensation. *Icarus*, **130**, 222–224.

Zhai, Y., Cummer, S.A., and Farrell, W.M. (2006). Quasi-electrostatic field analysis and simulation of martian and terrestrial dust. *Journal of Geophysical Research*, **111**, E06016, doi: 10.1029/2005JE002618.

Zharkov, V.N. (1996). The internal structure of Mars: a key to understanding the origin of terrestrial planets. *Solar System Research*, **30**, 456–465.

Zolotov, M.Y. and Shock, E.L. (2000). An abiotic origin for hydrocarbons in the Allan Hills 84001 martian meteorite through cooling of magmatic and impact-generated gases. *Meteoritics and Planetary Science*, **35**, 629–638.

Zuber, M.T. (2001). The crust and mantle of Mars. *Nature*, **412**, 220–227.

Zuber, M.T., Smith, D.E., Solomon, S.C., *et al.* (1998). Observations of the north polar region of Mars from the Mars Orbiter Laser Altimeter. *Science*, **282**, 2053–2060.

Zuber, M.T., Solomon, S.C., Phillips, R.J., *et al.* (2000). Internal structure and early thermal evolution of Mars from Mars Global Surveyor topography and gravity. *Science*, **287**, 1788–1793.

Zurek, R. and Martin, L. (1993). Interannual variability of planet-encircling dust storms on Mars. *Journal of Geophysical Research*, **98**, 3247–3259.

# Appendix: Mission reports

## Early missions

A summary of the early Mars missions can be found in chapter 1, Spacecraft Exploration of Mars, by Conway Snyder and Vassili Moroz (1992) in *Mars*, eds. H.H. Kieffer, B.M. Jakosky, C.W. Snyder, and M.S. Matthews. Tucson, AZ: University of Arizona Press, pp.71–119.

## Viking

Viking mission results were published in four special issues of the *Journal of Geophysical Research*:

Volume 82, 30 September 1977
Volume 84, 30 December 1979
Volume 87, 30 November 1982
Volume 95, 30 August 1990

## Mars Pathfinder

The Mars Pathfinder 90 day report was published in the 5 December 1997 issue of *Science* (vol. 278).
Results from the Mars Pathfinder mission were published in two issues of the *Journal of Geophysical Research*:

Volume 104, 25 April 1999
Volume 105, 25 January 2000

## Mars Global Surveyor

The Mars Global Surveyor 90 day report was published in the 13 March 1998 issue of *Science* (vol. 279).
A special section devoted to the Mars Global Surveyor mission initial results was published in the 25 October 2001, issue of the *Journal of Geophysical Research* (vol. 106).

## Mars Odyssey

Initial reports from Mars Odyssey can be found in the July 2002 (vol. 297) and June 2003 (vol. 300) issues of *Science*.

A special section devoted to the Mars Odyssey mission early results was published in the January 2004 issue of *Space Science Reviews* (vol. 110).

## Mars Express

A book describing the scientific experiments on Mars Express is: *Mars Express: The Scientific Payload*, edited by Andrew Wilson and published by ESA Publication Division (2004).

First results from the OMEGA instrument were published in the March 2005 issue of *Science* (vol. 307).

First results from the PFS instrument were published in the August 2005 issue of *Planetary and Space Science* (vol. 53).

## Mars Exploration Rovers

The 90 day report for the Spirit rover was published in the 6 August 2004, issue of *Science* (vol. 305).

The 90 day report for the Opportunity rover was published in the 3 December 2004, issue of *Science* (vol. 306).

## Mission Websites

US Mars mission information can be accessed through the JPL Mars Exploration Program website: mars.jpl.nasa.gov/

Mars 96: www.iki.rssi.ru/mars96/mars96hp.html

Mars Express: www.esa.int/SPECIALS/Mars.Express/index.html

ExoMars: www.esa.int/SPECIALS/Aurora/SEM1NVZKQAD.0.html

Nozomi: nssdc.gsfc.nasa.gov/nmc/tmp/1998-041A.html

www.stp.isas.jaxa.jp/nozomi/index-e.html

# Index

Printed in the United States
By Bookmasters